BIO-INSPIRED SMART HYDROGELS

仿生智能水凝胶

李明 赵润 李维军 管晴雯 编著

化学工业出版社
·北京·

内 容 简 介

本书从构建水凝胶的常见高分子着手，系统阐述了水凝胶的主要合成方法、内部的主要交联类型以及高分子网络的结构特征；详细介绍了功能性水凝胶力学性能和物理特性的设计原理以及相应的调控方法；进一步展开介绍了受自然界中刺激-响应驱动行为启发而构建的仿生智能水凝胶软执行器及其应用；特别关注了近几年兴起的自供能水凝胶传感器，详细阐述了相应的供能机制和应用领域；最后简要说明了该领域当前的挑战以及未来的发展方向。

本书可供材料、化学化工、机械、生物科技、人工智能等专业领域，尤其是仿生功能材料、响应性水凝胶、柔性传感器、可穿戴设备等研究领域的人员参考，亦可作为高等院校相关专业的教材，还可作为仿生爱好者的科普读物。

图书在版编目（CIP）数据

仿生智能水凝胶 / 李明等编著. —北京：化学工业出版社，2023.11
ISBN 978-7-122-44213-0

Ⅰ.①仿⋯ Ⅱ.①李⋯ Ⅲ.①水凝胶-仿生材料-智能材料-研究 Ⅳ.①TQ436

中国国家版本馆 CIP 数据核字（2023）第 179619 号

责任编辑：王清颢　　　　　　　　装帧设计：王晓宇
责任校对：李　爽

出版发行：化学工业出版社（北京市东城区青年湖南街13号　邮政编码100011）
印　　装：北京建宏印刷有限公司
710mm×1000mm　1/16　印张15　字数270千字　2024年1月北京第1版第1次印刷

购书咨询：010-64518888　　　　　　　售后服务：010-64518899
网　　址：http://www.cip.com.cn

凡购买本书，如有缺损质量问题，本社销售中心负责调换。

定　　价：158.00元　　　　　　　　　　　　　　　　版权所有　违者必究

编写人员名单

主要编写人员　李　明　赵　润　李维军　管晴雯
　　　　　　　李　昶　周诗桐

其他参编人员　Eduardo Saiz　王玉萍　侯　旭
　　　　　　　于志一　吕　静　王志航　丁　兰
　　　　　　　董潇潇

前言
PREFACE

水凝胶 (hydrogel) 是一种 3D 亲水交联聚合物网络，其物理特性类似于软生物组织，可以通过表面张力或毛细管效应容纳大量的水。自然界中，生物体的形态、模式和结构为科学的创新提供了灵感，也为解决仿生智能水凝胶在应用面临的问题上提供了宝贵的借鉴。从基础研究和实际应用的角度来讲，探索新型仿生智能水凝胶具有重要的现实意义。

本书以智能水凝胶为主线，借鉴仿生学思维，以介绍和讨论仿生水凝胶研究的科学策略，将其内部的构建机制应用到人造材料中解决实际问题为目标。仿生智能水凝胶具有可调的力学性能、优异的加工性能、自修复性能、多重黏附性等优异特性，从学术研究到工业领域的应用（包括药物输送、组织工程、医疗植入物和伤口敷料，以及传感器、软执行器、电子皮肤和柔性机器人等）都引起了人们的广泛关注。

尽管在过去这些年中，研究人员们已经开发出了大量的水凝胶并将其应用于不同的领域，但依然缺乏一套用于指导使用不同材料或制造方法来设计满足不同应用目的的水凝胶的通用原则，这也是本书要解决的核心问题。

本书从构建水凝胶的常见高分子着手，系统阐述了水凝胶的主要合成方法、内部的主要交联类型以及高分子网络的结构特征，详细介绍了功能性水凝胶力学性能和物理特性的设计原理以及相应的调控方法（本书中所介绍的这些设计原则和实施策略都是基于通用高分子网络的，同时，它们也适用于其他软材料，包括弹性体和有机凝胶）；随后，进一步展开介绍了受自然界中刺激－响应驱动行为启发而构建的仿生智能水凝胶软执行器及其应用；特别关注了近几年兴起的自供能水凝胶传感器，详细阐述了相应的供能机制和应用领域；最后简要说明了该领域当前的挑战以及未来的发展方向。

本书可供仿生、高分子、水凝胶、软体机器人、柔性传感器等研究领域人员参考，亦可作为相关领域爱好者的科普读物；本书涉及跨领域知识，还可作为高等院校材料类专业、化学化工类专业、机械专业、生物科技类专业等相关专业本科生或研究生的进阶教材，部分章节的内容适合电子、医学、能源和环境等相关领域特定二级学科学生学习了解。

Yu Shrike Zhang、高雪芹、赵选贺、Erica L.Bakota、David A.Tirrell、Francis G.Spinale、Partrick H.Campbell、陈俊、聂双喜、陈宝东等人为本书提供了大力支持，在此表示衷心的感谢。

由于编著者水平有限，书中难免会有疏漏之处，敬请读者批评指正。

编著者
2023 年 4 月

目录

第 1 章　绪论　001
参考文献　007

第 2 章　构建水凝胶的常见高分子　012
2.1　天然高分子　013
2.1.1　多糖类　013
2.1.2　蛋白质类　016
2.1.3　多肽类　017
2.1.4　核酸类　019
2.2　合成高分子　020
2.2.1　聚丙烯及其衍生物类　020
2.2.2　聚醇类　021
2.2.3　其他类　022
参考文献　023

第 3 章　水凝胶的合成方法　040
3.1　由温度引起的高分子链纠缠　041
3.2　分子自组装　042
3.3　离子凝胶化/静电相互作用　043

3.4 化学交联 044
3.5 小结 045
　　参考文献 045

第 4 章　水凝胶内部的主要交联类型　　048

4.1 永久共价交联 049
　　4.1.1 碳-碳键 050
　　4.1.2 碳-氮键 050
　　4.1.3 碳-氧键 051
　　4.1.4 碳-硫键 051
　　4.1.5 硅-氧键 051
4.2 强物理交联 051
　　4.2.1 晶畴 052
　　4.2.2 玻璃状结节 052
　　4.2.3 螺旋关联 053
4.3 弱物理交联 053
　　4.3.1 氢键 053
　　4.3.2 静电相互作用 054
　　4.3.3 配位络合 055
　　4.3.4 主客体相互作用 055
　　4.3.5 疏水缔合 056
　　4.3.6 π-π 堆积 057
4.4 动态共价交联 057
　　4.4.1 亚胺键 058
　　4.4.2 硼酸酯键 058
　　4.4.3 二硫键 059
　　4.4.4 腙键 059
　　4.4.5 肟键 060
　　4.4.6 可逆 Diels-Alder 反应 060

参考文献　　　　　　　　　　　　　　　　　061

第 5 章　水凝胶高分子网络的结构特征　　　　　077

5.1　弹性体水凝胶　　　　　　　　　　　　　　078
5.1.1　干燥状态下的弹性高分子网络　　　　　078
5.1.2　溶胀状态下的弹性高分子网络　　　　　080

5.2　非弹性体水凝胶　　　　　　　　　　　　　082
5.2.1　理想高分子网络　　　　　　　　　　　083
5.2.2　含有滑动交联点的高分子网络　　　　　083
5.2.3　互穿和半互穿高分子网络　　　　　　　084
5.2.4　具有高官能交联的高分子网络　　　　　085
5.2.5　微纳纤维高分子网络　　　　　　　　　085
5.2.6　其他非常规高分子网络　　　　　　　　086

5.3　由非常规高分子网络结构引起的力学性能分离　086

5.4　非常规高分子网络结构和相互作用的协同效应　089

参考文献　　　　　　　　　　　　　　　　　090

第 6 章　水凝胶极限力学性能的设计原理和调控方法　097

6.1　韧性：在可拉伸高分子网络中引入能量耗散机制　098
6.1.1　断裂韧性　　　　　　　　　　　　　　098
6.1.2　坚韧水凝胶的设计原则　　　　　　　　099
6.1.3　坚韧水凝胶的实施策略　　　　　　　　100

6.2　强度：让高分子网络内部有足够多的分子链能够同时硬化且断裂　　　　　　　　　　　　　　　　103
6.2.1　抗拉强度　　　　　　　　　　　　　　103
6.2.2　抗拉伸水凝胶的设计原则　　　　　　　104
6.2.3　抗拉伸水凝胶的实施策略　　　　　　　105

6.3　弹性：降低水凝胶在一定变形范围内的机械耗散　107

6.3.1	弹性	107
6.3.2	高弹性水凝胶的设计原则	107
6.3.3	高弹性水凝胶的实施策略	109

6.4 韧性黏结：整合具有机械耗散的增韧水凝胶基体与高强界面的交联 111

- 6.4.1 界面韧性 111
- 6.4.2 强界面黏附性水凝胶的设计原则 112
- 6.4.3 强界面黏附性水凝胶的实施策略 113

6.5 抗疲劳：用具有高本征断裂能的物质去阻碍疲劳裂纹扩展 115

- 6.5.1 疲劳阈值 115
- 6.5.2 抗疲劳水凝胶的设计原则 116
- 6.5.3 抗疲劳水凝胶的实施策略 117

6.6 抗疲劳粘接：在界面处强力固定具有高本征断裂能的物质 119

- 6.6.1 界面疲劳阈值 119
- 6.6.2 抗疲劳黏附水凝胶的设计原则 120
- 6.6.3 抗疲劳黏附水凝胶的实施策略 121

参考文献 122

第7章 水凝胶功能特性的设计原理和调控方法 130

- 7.1 导电性：形成连通的电子导电相 131
- 7.2 磁性：嵌入磁性颗粒并形成铁磁磁畴 132
- 7.3 折射率和透明度：均匀嵌入高折射率且无散射的纳米相 133
- 7.4 可调控声阻抗：等效均质水凝胶的密度和体积模量的调控 133
- 7.5 自愈性：在损伤区域形成新的交联或高分子链 134
- 7.6 可注射性：选择具有剪切变稀和自我修复特性的材料 136

参考文献 137

第8章 水凝胶的动态模拟 141

8.1 光图案化和光化降解法 142
8.2 动态光度图形法 143
8.3 细胞响应反馈系统法 144
8.4 刺激响应——形态变形法 145
8.5 细胞介导牵引力引起的形态变形法 146
  参考文献 147

第9章 仿生智能水凝胶软执行器及其应用 150

9.1 自然界中的刺激-响应驱动行为 151
 9.1.1 基于细胞渗透压变化实现的驱动行为 152
 9.1.2 基于纤维素原纤维结构不均匀膨胀实现的驱动行为 153
 9.1.3 基于可逆弱键的断裂/生成实现的驱动行为 153
 9.1.4 基于微观结构变化实现的驱动行为 154
 9.1.5 基于软结构的收缩/拉伸实现的驱动行为 155
9.2 人造刺激-响应性水凝胶执行器 155
 9.2.1 热响应 156
 9.2.2 光响应 161
 9.2.3 磁响应 165
 9.2.4 电响应 167
 9.2.5 pH响应 170
 9.2.6 离子响应 172
 9.2.7 湿度响应 173
 9.2.8 溶剂响应 175
 9.2.9 其他响应 176
9.3 仿生智能水凝胶执行器的应用 178

9.3.1	软执行器	179
9.3.2	流体操控	187
9.3.3	医学工程	191
参考文献		195

第 10 章　仿生自供能水凝胶传感器　205

10.1　自供能水凝胶传感器的供能机制　206

10.1.1	摩擦纳米发电	207
10.1.2	压电纳米发电	208
10.1.3	热电纳米发电	209
10.1.4	光伏发电	209
10.1.5	水伏发电	210
10.1.6	磁电发电	211
10.1.7	混合发电	211

10.2　自供能水凝胶传感器的典型应用　212

10.2.1	物理传感	212
10.2.2	健康护理	215
10.2.3	环境监测	217
参考文献		219

第 11 章　总结与展望　222

11.1　仿生智能水凝胶软执行器　223

11.2　自供能水凝胶传感器　224

第 1 章
绪论

参考文献

水凝胶（hydrogel）是以水为分散介质的凝胶。具有三维网状结构的水溶性高分子材料中引入一部分疏水基团和亲水基团，亲水基团可结合水分子，将其连接在网状内部，而疏水基团是一种遇水膨胀的交联聚合物。水凝胶，既是高分子的浓溶液，又是高弹性的固体。水分子在聚合物网络中以键合水、束缚水和游离水等形式存在而失去了流动性，水溶性物质可以在其内部渗透或扩散。

作为一种含有大量水的亲水性高分子网络，水凝胶是动物体内的主要构成成分，包含了绝大部分的细胞、细胞外基质、组织和器官[1]。由于其优异的生物相容性，水凝胶已广泛用于生物医学中（图1.1），例如药物输送载体[2-5]、组织工程支架[6-8]、医疗植入物[9,10]、伤口敷料[11-13]和隐形眼镜等[10,14]。

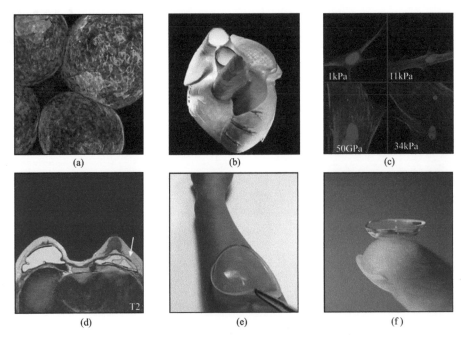

图1.1 水凝胶典型的生物医学应用[15]
（a）药物输送载体；（b）组织工程支架；（c）生物学研究模型；
（d）医疗植入物；（e）伤口敷料；（f）隐形眼镜

近年来，研究人员致力于探索水凝胶在设备和机器中的新兴应用（图1.2）[16]。例如水凝胶传感器[17-20]、软执行器[21-24]、柔性机器人[22,25,26]、柔性电池[27,28]、水凝胶超级电容器[29]、离子电子设备[30]、磁性设备[31-33]、光学设备[34-36]、声学设备[37]、水下黏合剂[38-40]、生物黏合剂[38,41]，以及涂层[42,43]等。

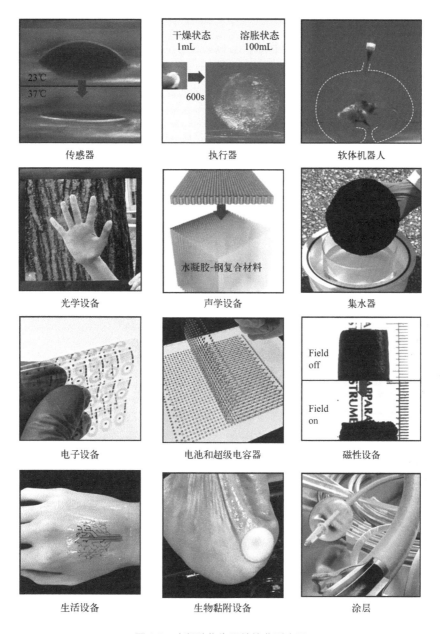

图1.2 水凝胶作为器件的典型应用

随着水凝胶研究的不断深入，水凝胶材料的力学性能对于动物的生存、健康以及其他功能应用至关重要。尽管在过去几十年中，高分子和软材料领域的开创性工作为理解水凝胶的弹性、膨胀、多孔弹性、黏弹性、断裂和疲劳奠定了基础[44-47]。然而，如何设计具有某些优异力学性能或某些特殊功能特性的水凝胶，仍然是高分子和软材料领域的一个巨大挑战[48-51]。当人们以水凝胶的优异力学性能为目标时，这一挑战变得更加艰巨，例如优异的断裂韧性值[52]、强度[53,54]、弹性[55,56]、界面韧性[40]、疲劳阈值[57-59]以及界面疲劳阈值[60]。虽然现如今的研究仍面临上述巨大挑战，但具有优异力学性能的水凝胶的设计是有一定的应用基础的。这是因为许多生物水凝胶已经通过进化获得了生存所必需的优异力学性能并且应用到了其日常的生理活动之中（图1.3），这可以为我们设计相应的具有优异力学性能的人工水凝胶提供灵感。

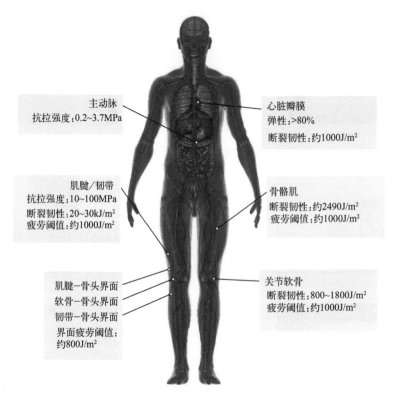

图1.3　人体内的生物水凝胶的力学性能[15]

从人体结构分析，水凝胶材料发挥着重要的作用。

软骨，作为一种坚韧的结缔组织，覆盖在关节表面以减少摩擦[61]。人膝关节软骨（即关节软骨）每年通常需要承受约100万次4～9MPa的压缩应力循环，同时保持约1000J/m²的高断裂韧性（图1.4）[62]。关节软骨的高断裂韧性主要归因于其内部丰富的强胶原纤维与蛋白多糖大分子之间的相互渗透所形成的多孔结构，其结构为机械耗散提供了黏弹性和多孔弹性[63,64]。关节软骨的黏弹性主要与聚集蛋白多糖的局部重排和黏附相互作用以及胶原蛋白的重构有关[64]；而关节软骨的多孔弹性则受多孔细胞外基质中间质液的运动控制[63]。

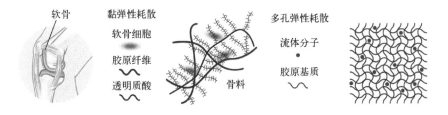

图 1.4　由于聚合物网络的黏弹性和多孔弹性耗散，软骨具有高韧性[15]

肌腱是一种强壮的结缔组织，可将肌肉与骨骼以及肌肉与肌肉连接起来。人类的髌腱可以维持超过 **50MPa** 的高拉伸强度[65,66]，这是由于其独特的分级纤维结构，能够在胶原纤维束拉伸失效之前硬化（图 1.5）[67,68]。

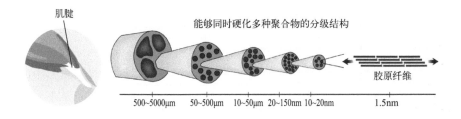

图 1.5　由于纤维分级结构中多种聚合物的同时硬化，肌腱具有高拉伸强度[15]

心脏瓣膜通常同时具有 **80%** 以上的高弹性和 **1200J/m²** 左右的高断裂韧性[69,70]，见图 1.6。由于心脏瓣膜中的弹性蛋白和卷曲胶原纤维在适度变形下具有弹性和非耗散性，因此心脏瓣膜具有高弹性[71]。形变状态下，胶原纤维的硬化和断裂会消耗大量的机械能，从而赋予心脏瓣膜高韧性[72]。

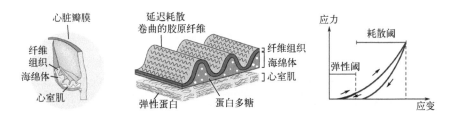

图 1.6　由于延迟的机械耗散，心脏瓣膜具有高弹性和高断裂韧性[15]

软结缔组织在骨骼上的黏附可以极度抗疲劳。例如，人膝关节中的软骨-骨界面可以承受 1MPa 的压力以及大约 800J/m² 的界面韧性，这种载荷循环每年超过 100 万次（图 1.7）[62,73,74]。进一步的研究发现，软组织（例如，肌腱、韧带和软骨）与刚性骨骼的抗疲劳黏附通常是通过由胶原纳米纤维和有序的羟基磷灰石纳米晶体组成的纳米结构界面实现的[75-77]。

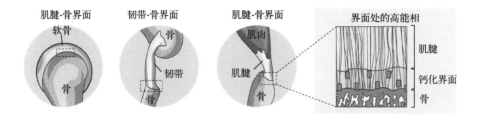

图1.7　由于在界面上固有的高能相（包含了牢固结合的纳米晶体和纳米纤维），软骨／韧带／肌腱－骨界面具有高界面疲劳阈值[15]

受到这些生物水凝胶优异力学性能的启发，在过去的几十年里，科研人员使用大量的候选材料（包括各种天然和合成聚合物，纳米／微米／大分子填料，以及纳米／微米／大分子纤维）尝试开发具有极端力学性能的水凝胶。尽管其中一些水凝胶的性能非常出色，但它们的设计通常遵循爱迪生的方法——对特定候选材料进行反复试验。随着该领域的快速发展，水凝胶在生物医学和机器中的新兴应用对水凝胶的合理引导设计提出了更高的要求，需要人们可以从不同的候选材料和制造方法中进行选择来设计水凝胶，从而实现目标的力学性能。然而，各种生物水凝胶实现优异力学性能的自然设计原则究竟是什么，仍是一个悬而未决的问题。

水凝胶的应用通常需要其具有适应于应用场景的特定性能。例如，用于调节干细胞的水凝胶需要具有不同模量和黏弹性[75,78-80]，作为人造软骨和椎间盘应用的水凝胶需要在循环机械载荷下具有抗疲劳性[57,58,81,82]，应用于受控药物输送领域的水凝胶需要精确调控其高分子网络的网眼大小[35,83,84]。此外，水凝胶作为各种设备和功能器件的载体，同样需要具有某种／某些特定的功能特性。例如，水凝胶传感器和执行器需要对外部刺激具有响应能力[26,85-87]，水凝胶涂层需要具有强黏附性[40]，水凝胶光学器件需具有优异的透明度[35]，水凝胶电子器件需具有导电性[88]，水凝胶集水器则需要具有相应的吸水／释放水的性能[89]。

因此，本书旨在：
① 阐述用于设计水凝胶以获得极端力学性能的通用原则，包括极高的断裂韧性、拉伸强度、回弹性、界面韧性、疲劳阈值和界面疲劳阈值；
② 分析构建极端的物理特性的通用原则，包括高导电性、图案化磁化、高折射率和透明度、可调声阻抗和自我修复；
③ 介绍基于非常规聚合物网络（UPN）的构建策略；

④ 介绍用于新型水凝胶设计和制造的正交设计原则和协同实施策略，以实现多种组合的力学、物理、化学和生物特性；

⑤ 介绍基于书中所述设计原则和实施策略所构建出的仿生水凝胶器件以及其在智能驱动和柔性传感领域的应用。

参考文献

[1] Wichterle O, Lim D. Hydrophilic gels for biological use. Nature, 1960, 185 (4706): 117-118.

[2] Peppas N A, Bures P, Leobandung W, et al. Hydrogels in pharmaceutical formulations. European journal of pharmaceutics and biopharmaceutics, 2000, 50 (1): 27-46.

[3] Qiu Y, Park K. Environment-sensitive hydrogels for drug delivery. Advanced drug delivery reviews, 2001, 53 (3): 321-339.

[4] Peppas N A, Hilt J Z, Khademhosseini A, et al. Hydrogels in biology and medicine: from molecular principles to bionanotechnology. Advanced materials,2006, 18 (11): 1345-1360.

[5] Li J, Mooney D J. Designing hydrogels for controlled drug delivery. Nature Reviews Materials, 2016, 1 (12): 1-17.

[6] Lee K Y, Mooney D J. Hydrogels for tissue engineering. Chemical reviews, 2001, 101 (7): 1869-1880.

[7] Nguyen K T, West J L. Photopolymerizable hydrogels for tissue engineering applications. Biomaterials, 2002, 23 (22): 4307-4314.

[8] Drury J L, Mooney D J. Hydrogels for tissue engineering: scaffold design variables and applications. Biomaterials, 2003, 24 (24): 4337-4351.

[9] Caló E, Khutoryanskiy V V. Biomedical applications of hydrogels: A review of patents and commercial products. European polymer journal, 2015, 65: 252-267.

[10] Kopecek J. Hydrogels: From soft contact lenses and implants to self-assembled nanomaterials. Journal of Polymer Science Part A: Polymer Chemistry, 2009, 47 (22): 5929-5946.

[11] Jones A, Vaughan D. Hydrogel dressings in the management of a variety of wound types: A review. Journal of Orthopaedic Nursing, 2005, 9: S1-S11.

[12] Boateng J S, Matthews K H, Stevens H N, et al. Wound healing dressings and drug delivery systems: a review. Journal of pharmaceutical sciences, 2008, 97 (8): 2892-2923.

[13] Kamoun E A, Kenawy E R S, Chen X. A review on polymeric hydrogel membranes for wound dressing applications: PVA-based hydrogel dressings. Journal of advanced research, 2017, 8 (3): 217-233.

[14] Vanderlaan D G, Turner D C, Hargiss M V, et al. Soft contact lenses. Google Patents, 2005.

[15] Zhao X, Chen X, Yuk H, et al. Soft materials by design: unconventional polymer networks give extreme properties. Chemical Reviews, 2021, 121 (8): 4309-4372.

[16] Liu X, Liu J, Lin S, et al. Hydrogel machines. Materials Today, 2020, 36: 102-124.

[17] Dong L, Agarwal A K, Beebe D J, et al. Adaptive liquid microlenses activated by stimuli-responsive hydrogels. Nature, 2006, 442 (7102): 551-554.

[18] Holtz J H, Asher S A. Polymerized colloidal crystal hydrogel films as intelligent chemical sensing materials. Nature, 1997, 389 (6653): 829-832.

[19] Miyata T, Asami N. Uragami T. A reversibly antigen-responsive hydrogel. Nature, 1999, 399 (6738): 766-769.

[20] Liu X, Tang T C, Tham E, et al. Stretchable living materials and devices with hydrogel-elastomer hybrids hosting programmed cells. Proceedings of the National Academy of Sciences, 2017, 114 (9): 2200-2205.

[21] Beebe D J, Moore J S, Bauer J M, et al. Functional hydrogel structures for autonomous flow control inside microfluidic channels. Nature, 2000, 404 (6778): 588-590.

[22] Yuk H, Lin S, Ma C, et al. Hydraulic hydrogel actuators and robots optically and sonically camouflaged in water. Nature communications, 2017, 8 (1): 14230.

[23] Suo Z. Stretchable, transparent, ionic conductors. Science, 341 (6149): 984.

[24] Zhao X, Kim J, Cezar C A, et al. Active scaffolds for on-demand drug and cell delivery. Proceedings of the National Academy of Sciences, 2011, 108 (1): 67-72.

[25] Li T, Li G, Liang Y, et al. Fast-moving soft electronic fish. Science advances, 2017, 3 (4): e1602045.

[26] Kim Y S, Liu M, Ishida Y, et al. Thermoresponsive actuation enabled by permittivity switching in an electrostatically anisotropic hydrogel. Nature materials, 2015, 14 (10): 1002-1007.

[27] Schroeder T B, Guha A, Lamoureux A, et al. An electric-eel-inspired soft power source from stacked hydrogels. Nature, 2017, 552 (7684): 214-218.

[28] Pan L, Yu G, Zhai D, et al. Hierarchical nanostructured conducting polymer hydrogel with high electrochemical activity. Proceedings of the National Academy of Sciences, 2012, 109 (24): 9287-9292.

[29] Xiao T, Xu L, Zhou L, et al. Dynamic hydrogels mediated by macrocyclic host-guest interactions. Journal of Materials Chemistry B, 2019, 7 (10): 1526-1540.

[30] Kim C C, Lee H H, Oh K H, et al. Highly stretchable, transparent ionic touch panel. Science, 2016, 353 (6300): 682-687.

[31] Kim Y, Parada G A, Liu S, et al. Ferromagnetic soft continuum robots. Science Robotics, 2019, 4 (33): eaax7329.

[32] Kim Y, Yuk H, Zhao R, et al. Printing ferromagnetic domains for untethered fast-transforming soft materials. Nature, 2018, 558 (7709): 274-279.

[33] Wang L, Kim Y, Guo C F, et al. Hard-magnetic elastica. Journal of the mechanics and physics of solids, 2020, 142: 104045.

[34] Chen F, Tillberg P W, Boyden E S. Expansion microscopy. Science, 2015, 347 (6221): 543-548.

[35] Guo J, Liu X, Jiang N, et al. Highly stretchable, strain sensing hydrogel optical fibers. Advanced Materials, 2016, 28 (46): 10244-10249.

[36] Li X H, Liu C, Feng S P, et al. Broadband light management with thermochromic hydrogel microparticles for smart windows. Joule, 2019, 3 (1): 290-302.

[37] Liu X, Yuk H, Lin S, et al. 3D printing of living responsive materials and devices. Advanced Materials, 2018, 30 (4): 1704821.

[38] Yuk H, Varela C E, Nabzdyk C S, et al. Dry double-sided tape for adhesion of wet tissues and devices. Nature, 2019, 575 (7781): 169-174.

[39] Rose S, Prevoteau A, Elzière P, et al. Nanoparticle solutions as adhesives for gels and biological tissues. Nature, 2014, 505 (7483): 382-385.

[40] Yuk H, Zhang T, Lin S, et al. Tough bonding of hydrogels to diverse non-porous surfaces. Nature materials, 2016, 15 (2): 190-196.

[41] Li J, Celiz A, Yang J, et al. Tough adhesives for diverse wet surfaces. Science, 2017, 357 (6349): 378-381.

[42] Yu Y, Yuk H, Parada G A, et al. Multifunctional "hydrogel skins" on diverse polymers with arbitrary shapes. Advanced Materials, 2019, 31 (7): 1807101.

[43] Cheng H, Yue K, Kazemzadeh-Narbat M, et al. Mussel-inspired multifunctional hydrogel coating for prevention of infections and enhanced osteogenesis. ACS applied materials & interfaces, 2017, 9 (13): 11428-11439.

[44] Biot M A, General theory of three-dimensional consolidation. Journal of applied physics, 1941, 12 (2): 155-164.

[45] Flory P J. Principles of polymer chemistry. Cornell university press, 1953.

[46] Lake G, Lindley P. The mechanical fatigue limit for rubber. Journal of Applied Polymer Science, 1965, 9 (4): 1233-1251.

[47] Lake G, Thomas A. The strength of highly elastic materials. Proceedings of the Royal Society of London Series A Mathematical and Physical Sciences, 1967, 300 (1460): 108-119.

[48] Richbourg N R, Peppas N A. The swollen polymer network hypothesis: Quantitative models of hydrogel swelling, stiffness, and solute transport. Progress in Polymer Science, 2020, 105: 101243.

[49] Zhao X. Multi-scale multi-mechanism design of tough hydrogels: building dissipation into stretchy networks. Soft matter, 2014, 10 (5): 672-687.

[50] Zhang Y S, Khademhosseini A. Advances in engineering hydrogels. Science, 2017, 356 (6337): eaaf3627.

[51] Fan H, Gong J P. Fabrication of bioinspired hydrogels: challenges and opportunities. Macromolecules, 2020, 53 (8): 2769-2782.

[52] Gong J P, Katsuyama Y, Kurokawa T, et al. Double-network hydrogels with extremely high mechanical strength. Advanced materials, 2003, 15 (14): 1155-1158.

[53] Moutos F T, Freed L E, Guilak F. A biomimetic three-dimensional woven composite scaffold for functional tissue engineering of cartilage. Nature materials, 2007, 6 (2): 162-167.

[54] Hua J, Ng P F, Fei B. High-strength hydrogels: Microstructure design, characterization and applications. Journal of Polymer Science Part B: Polymer Physics, 2018, 56 (19): 1325-1335.

[55] Lin S, Zhou Y, Zhao X. Designing extremely resilient and tough hydrogels via delayed dissipation. Extreme Mechanics Letters, 2014, 1: 70-75.

[56] Kamata H, Akagi Y, Kayasuga-Kariya Y, et al. "Nonswellable" hydrogel without mechanical hysteresis. Science, 2014, 343 (6173): 873-875.

[57] Lin S, Liu X, Liu J, et al. Anti-fatigue-fracture hydrogels. Science advances, 2019, 5 (1): eaau8528.

[58] Lin S, Liu J, Liu X, et al. Muscle-like fatigue-resistant hydrogels by mechanical training. Proceedings of the National Academy of Sciences, 2019, 116 (21): 10244-10249.

[59] Xiang C, Wang Z, Yang C, et al. Stretchable and fatigue-resistant materials. Materials Today, 2020, 34: 7-16.

[60] Liu J, Lin S, Liu X, et al. Fatigue-resistant adhesion of hydrogels. Nature communications, 2020, 11 (1): 1071.

[61] Hall, B. K., Cartilage V1: Structure, function, and biochemistry. Academic Press: 2012.

[62] Taylor D, O'Mara N, Ryan E, et al. The fracture toughness of soft tissues. Journal of the mechanical behavior of biomedical materials, 2012, 6: 139-147.

[63] Mow V C, Holmes M H, Lai W M, Fluid transport and mechanical properties of articular cartilage: a review. Journal of biomechanics, 1984, 17 (5): 377-394.

[64] Han L, Frank E H, Greene J J, et al. Time-dependent nanomechanics of cartilage. Biophysical journal, 2011, 100 (7): 1846-1854.

[65] Tavichakorntrakool R, Prasongwattana V, Sriboonlue P, et al. Chamsuwan, A., K^+, Na^+, Mg^{2+}, Ca^{2+}, and water contents in human skeletal muscle: correlations among these monovalent and divalent cations and their alterations in K^+-depleted subjects. Translational research, 2007, 150 (6): 357-366.

[66] Johnson G A, Tramaglini D M, Levine R E, et al. Tensile and viscoelastic properties of human patellar tendon. Journal of orthopaedic research, 1994, 12 (6): 796-803.

[67] Connizzo B K, Yannascoli S M, Soslowsky L J. Structure-function relationships of postnatal tendon development: A parallel to healing. Matrix Biology, 2013, 32 (2): 106-116.

[68] Fratzl P, Collagen: structure and mechanics, an introduction. Springer, 2008.

[69] Lee J M, Courtman D W, Boughner D R. The glutaraldehyde-stabilized porcine aortic valve xenograft I Tensile viscoelastic properties of the fresh leaflet material. Journal of Biomedical Materials Research, 1984, 18 (1): 61-77.

[70] Lee J M, Boughner D R, Courtman, D W. The glutaraldehyde-stabilized porcine aortic valve xenograft II Effect of fixation with or without pressure on the tensile viscoelastic properties of the leaflet material. Journal of biomedical materials research, 1984, 18 (1): 79-98.

[71] Vesely I. The role of elastin in aortic valve mechanics. Journal of biomechanics, 1997, 31 (2): 115-123.

[72] Driessen N, J, Bouten C V, Baaijens F P. Improved prediction of the collagen fiber architecture in the aortic heart valve. 2005.

[73] Bobyn J, Wilson G, MacGregor D, et al. Effect of pore size on the peel strength of attachment of fibrous tissue to porous-surfaced implants. Journal of biomedical materials research, 1982, 16 (5): 571-584.

[74] Covert R J, Ott R, Ku D N. Friction characteristics of a potential articular cartilage biomaterial. Wear, 2003, 255 (7-12): 1064-1068.

[75] Chaudhuri O, Gu L, Klumpers D, et al. Hydrogels with tunable stress relaxation regulate stem cell fate and activity. Nature materials, 2016, 15 (3): 326-334.

[76] Genin G M, Thomopoulos S. Unification through disarray. Nature materials, 2017, 16 (6): 607-608.

[77] Rossetti L, Kuntz L, Kunold E, et al. The microstructure and micromechanics of the tendon-bone insertion. Nature materials, 2017, 16 (6): 664-670.

[78] Engler A J, Sen S, Sweeney H L, et al. Matrix elasticity directs stem cell lineage specification. Cell,

2006, 126 (4): 677-689.

[79] Huebsch N, Arany P R, Mao A S, et al. Rivera-Feliciano, J.; Mooney, D. J., Harnessing traction-mediated manipulation of the cell/matrix interface to control stem-cell fate. Nature materials, 2010, 9 (6): 518-526.

[80] Chaudhuri O, Cooper-White J, Janmey P A, et al. Effects of extracellular matrix viscoelasticity on cellular behaviour. Nature, 2020, 584 (7822): 535-546.

[81] Weightman B, Freeman M, Swanson S. Fatigue of articular cartilage. Nature, 1973, 244 (5414): 303-304.

[82] Hansson T, Keller T, Spengler D. Mechanical behavior of the human lumbar spine. II. Fatigue strength during dynamic compressive loading. Journal of Orthopaedic Research, 1987, 5 (4): 479-487.

[83] Peppas N A. Hydrogels and drug delivery. Current opinion in colloid & interface science, 1997, 2 (5): 531-537.

[84] Peppas N A, Khare A R. Preparation, structure and diffusional behavior of hydrogels in controlled release. Advanced drug delivery reviews, 1993, 11 (1-2): 1-35.

[85] Lewis J, Nuzzo R, Mahadevan L, et al. Biomimetic 4D printing. Nat Mat, 2016, 15: 413-418.

[86] Yang J, Li Y, Zhu L, et al. Double network hydrogels with controlled shape deformation: A mini review. Journal of Polymer Science Part B: Polymer Physics, 2018, 56 (19): 1351-1362.

[87] Peng X, Wang H. Shape changing hydrogels and their applications as soft actuators. Journal of Polymer Science Part B: Polymer Physics, 2018, 56 (19): 1314-1324.

[88] Lu B, Yuk H, Lin S, et al. Pure pedot: Pss hydrogels. Nature communications, 2019, 10 (1): 1043.

[89] Zhao F, Zhou X, Shi Y, et al. Highly efficient solar vapour generation via hierarchically nanostructured gels. Nature nanotechnology, 2018, 13 (6): 489-495.

第 2 章
构建水凝胶的常见高分子

2.1 天然高分子
2.2 合成高分子
参考文献

大量的高分子和交联物已被用于各种水凝胶的设计和制造。这些高分子可按照来源分为天然高分子和合成高分子。其中，前者可进一步分为多糖类高分子、蛋白质类高分子、多肽类高分子，以及核酸类高分子；而后者可按照其聚合单体的种类进行划分。在本章中，我们将简要介绍制备水凝胶常用的天然高分子与合成高分子。这些高分子在水凝胶中的各种交联类型将在第4章中进行讨论。

2.1 天然高分子

人类社会的发展始终伴随着天然高分子材料的利用。天然衍生的高分子，具有完整而严谨的超分子体系，可应用于构建水凝胶聚合物网络。基于天然高分子的水凝胶，由于其与生物组织具有相似的化学组成，通常具有良好的生物相容性，可通过新陈代谢和组织重塑过程在生物体内降解并被吸收。此外，大多数天然高分子拥有易于交联和改性的反应位点，这可以赋予相应的水凝胶以定制的生物、化学以及力学性能。在本节中，我们将简要介绍几种常用于构建水凝胶的天然高分子。

2.1.1 多糖类

2.1.1.1 藻酸盐

藻酸盐，通常从褐藻细胞壁以及固氮菌或假单胞菌中获得[1]，是一类包含 β-$(1 \rightarrow 4)$-连接的 D-甘露糖醛酸（M）和 α-$(1 \rightarrow 4)$-连接的 L-古洛糖醛酸（G）残基嵌段的线型共聚物（图2.1）。这些嵌段通常由连续的 G 残基（GGGGGG），连续的 M 残基（MMMMMM），以及交替的 M 残基和 G 残基（GMGMGM）组成[2]。藻酸盐水凝胶可以通过各种共价交联和物理交联形成。其中，基于离子实现的交联已被广泛用于藻酸盐水凝胶的构建，因为藻酸盐中 G 嵌段[3]和 GM 嵌段[4]可以很容易地通过二价阳离子相互结合（例如 Ca^{2+}、Mg^{2+}、Ba^{2+} 和 Sr^{2+}）[5-7]。此外，通过改变不同的参数，如分子量、聚合物浓度、化学修饰、G/M 以及交联的类型或密度，可以调整藻酸盐水凝胶的力学性能以让其适应于各种生物组织[3,8]。由于藻酸盐形成的水凝胶可轻松封装细胞和药物，因而已作为支架材料广泛用于组织工程中，例如椎间盘再生[9]、脂肪组织再生[10]、心脏再生[11]和肝再生[12]等。

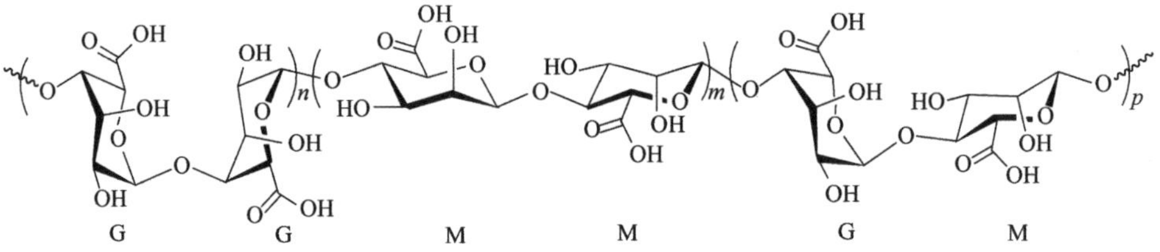

图 2.1　藻酸盐的化学结构示意图

2.1.1.2　琼脂糖

琼脂糖是由 β-D- 吡喃半乳糖和 3,6- 脱水 -α-L- 吡喃半乳糖组成的中性多糖（图 2.2），主要从红藻（红藻科）中提取[13]。作为一种典型的热敏性高分子，琼脂糖可以先通过加热溶解在水中，然后冷却形成水凝胶。胶凝过程中，琼脂糖的结构会从随机螺旋构型变为在连接区具有多链聚集的双螺旋束[14,15]。琼脂糖水凝胶的胶凝温度和力学性能可通过改变水凝胶中琼脂糖的浓度、分子量和结构来调节[16,17]。由于琼脂糖水凝胶在人体中的免疫反应较低[18]，因此可用于细胞封装[19]、软骨修复[20]和神经再生[21]等。值得注意的是，天然琼脂糖不具有细胞黏附基序，因此可以用共价交联的方式将细胞黏附肽结合到琼脂糖主链上，以增强细胞与琼脂糖水凝胶之间的相互作用[22]。

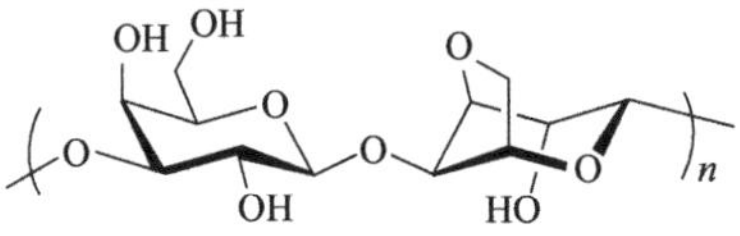

图 2.2　琼脂糖的化学结构示意图

2.1.1.3　壳聚糖

壳聚糖是由 β-（1 → 4）- 连接的 D- 氨基葡萄糖和 N- 乙酰 -D- 氨基葡萄糖组成的线型多糖（图 2.3）。其主要通过几丁质（从蟹壳和虾壳中获得）部分脱乙酰化获得，其中 N- 乙酰 -D- 氨基葡萄糖残基的含量少于 40%[23,24]。壳聚糖材料的物理、化学和生物学特性，与壳聚糖的分子量和脱乙酰度高度相关[25,26]。壳聚糖可以通过疏水相互作用、氢键[27,28]、金属配位 [Pt（Ⅱ）、Pd（Ⅱ）、Mo（Ⅳ）][29,30] 以及静电相互作用（硫酸根、柠檬酸根和磷酸根离子）[31,32] 与阴离子聚电解质（多糖[33,34]、蛋白质[35,36]和合成高分子[37]）形成物理交联的水凝胶。一般情况下，这些物理交联的壳聚糖水凝胶的力学性能低、使用寿命短，并且受 pH 值、温度和离子强度的影响也很大[24,38]。为了提高壳聚糖水凝胶的力学性能和稳定性，可以将共价交联剂引入壳聚糖水凝胶中。常用的共价交联剂包括二醛[39,40]、甲醛[41]、二缩水甘油醚[42]和京尼平[43,44]等，它们可以与壳聚糖主链上的残留官能团（如—OH、—COOH和—NH$_2$）发生反应形成酰胺键、酯键和席夫碱键等[27,45,46]。此外，甲基丙烯酸酯或芳基叠氮基团化学修饰的壳聚糖，还可通过光交联，形成大分子单体[47]。这类可光交联的壳聚糖水凝胶的胶凝度和力学性能可以通过紫外线照射时间和强度来控制[48-50]。壳聚糖水凝胶还可以用生物功能配体如Arg-Gly-Asp（RGD）肽进行修饰，以促进细胞黏附和增殖[51,52]。壳聚糖水凝胶，由于其出色的生物相容性和生物降解性[53]，在药物输送[54]、细胞封装[55]、神经组织工程[56]和骨再生[55]等生物医学领域具有极高应用价值。

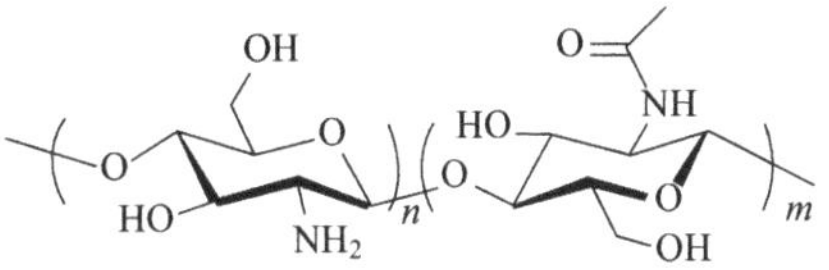

图2.3 壳聚糖的化学结构示意图

2.1.1.4 纤维素

纤维素，作为自然界中含量最丰富的天然多糖，是植物和棉麻等天然纤维的主要成分（图2.4）[57-59]。此外，一些细菌（例如木醋杆菌）也能产生纤维素[60]。纤维素由 β-（1→4）-连接的D-葡萄糖单元组成，因此纤维素具有高结晶度（超过40%），从而难溶于水和其他常用溶剂[61]。为提升天然纤维素的溶解性，丰富的水性溶剂体系如N-甲基吗啉-N-氧化物[62,63]、离子液体[64,65]和碱/尿素（或硫脲）[66,67]等已被开发应用。对纤维素主链的羟基部分酯化或醚化，可实现其功能改性[57]。这些改性后的纤维素衍生物，如甲基纤维素[68]、羟丙基纤维素[69]、羟丙基甲基纤维素[70,71]、羧甲基纤维素[72]等，比天然纤维素更加易于溶解和加工。

纤维素及其衍生物通过化学交联可形成稳定的三维网络。含双官能团或多官能团基的分子，如1,2,3,4-丁烷四甲酸二酐[73]、琥珀酸酐[74]、柠檬酸[75]、环氧氯丙烷[76]、乙二醇二缩水甘油醚[77]、二乙烯基砜[78]等，作为交联剂，可在纤维素链之间形成共价酯键或醚键。此外，纤维素链也可通过电子束和γ射线照射进行共价交联[79,80]，该方法不仅可以避免使用有毒交联剂，还可对所得水凝胶灭菌。纤维素及其衍生物通过与天然聚合物（如壳聚糖[81]、淀粉[82]、藻酸盐[83]、透明质酸[84]）以及合成聚合物（聚乙二醇[85]、聚乙烯醇[86]、聚N,N-二甲基丙烯酰胺[87]）混合，可形成具有优异力学性能的互穿聚合物网络。研究表明，由某些细菌物种（如木醋杆菌）产生的细菌纤维素可直接形成具有高纯度和拉伸强度的纤维素水凝胶[88,89]。纤维素基水凝胶优异的亲水性、生物降解性、生物相容性和透明性，使其在药物输送[90]、组织工程[91]、血液净化[92]、应变传感器[93]、净水[94]等领域具有应用前景。

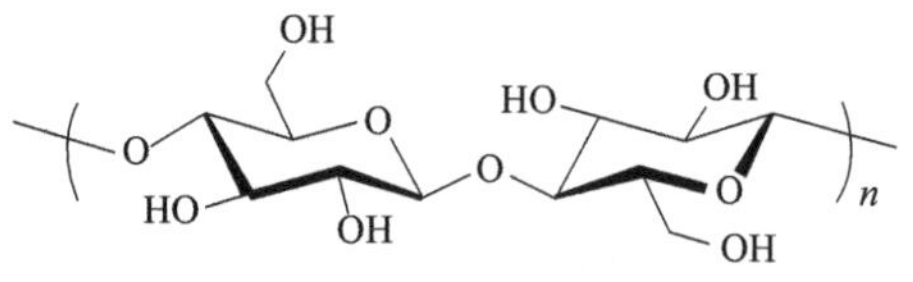

图 2.4　纤维素的化学结构示意图

2.1.2　蛋白质类

2.1.2.1　纤维蛋白

纤维蛋白是一种天然衍生的高分子，通常从凝血酶处理的纤维蛋白原中获得（图 2.5）[95]。纤维蛋白可通过形成纤维网络参与伤口的自然愈合过程。在室温下混合纤维蛋白原和凝血酶溶液时，纤维蛋白会形成凝块或水凝胶[96]。由物理交联形成的纤维蛋白水凝胶，其力学性能相对较低；为了改善纤维蛋白水凝胶的力学性能，可以引入京尼平等化学交联剂来交联纤维蛋白上的氨基残基，从而形成稳定的共价交联网络[97]。此外，纤维蛋白水凝胶还可以与聚氨酯[98]、聚己内酯[99]、β-磷酸三钙[100]、聚乙二醇[101]等人工合成高分子结合，以增强水凝胶的机械强度。纤维蛋白水凝胶可用作组织密封剂和黏合剂来控制手术中的出血[102]，并被广泛用于心脏组织工程的支架[103]、神经再生[104]、眼部治疗[105]、软骨和骨骼修复[106,107]、肌肉组织工程[108]和伤口愈合中的外源性递送等领域[109]。值得注意的是，纤维蛋白水凝胶可以从患者自己的血液中提取，因此在生物医学工程运用中可显著降低异物反应的风险[110]。

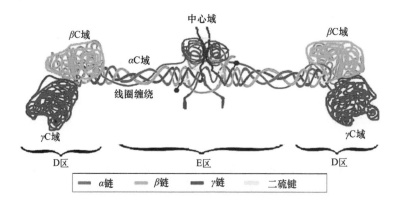

图 2.5　纤维蛋白的化学结构示意图[95]

2.1.2.2 胶原蛋白

胶原蛋白是动物体内的主要蛋白质之一，目前已经发现的胶原蛋白大约有29种[111]。胶原蛋白中存在多级结构（图2.6），分别为一级结构（氨基酸三联体）、二级结构（α-螺旋）、三级结构（三螺旋）和四级结构（原纤维）[112,113]。胶原蛋白的一级结构是—(Gly-X-Y-)$_n$—的三肽序列。其中，Gly是甘氨酸；X和Y是Gly以外的其他氨基酸。氨基酸的序列决定了肽折叠成二级结构的方式，二级结构以左手α-螺旋多肽链为主，通过氨基酸残基之间的氢键稳定[114]。三个左手α-螺旋多肽链可以通过羟醛缩合交联、醛胺缩合交联和羟醛组氨酸交联形成三级结构[115]，并进一步自组装成四级结构的胶原纤维[116]。

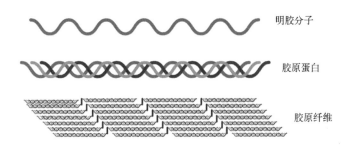

图2.6 胶原蛋白的化学结构示意图

当胶原溶液被加热中和时，酸溶胶原可以自组装形成物理交联的水凝胶。由于物理交联的胶原蛋白水凝胶通常力学性能较差且热不稳定[117,118]，因此需要进一步通过化学交联（如戊二醛、京尼平、碳二亚胺、二苯基磷酰叠氮等）来强化和稳定[119-121]。研究表明，胶原蛋白可以被胶原酶和金属蛋白酶生物降解，经交联后，其降解速率可显著降低[122]。由于胶原蛋白通常具有低抗原性、低炎症反应、良好的生物相容性和天然细胞黏附基序[123-125]，胶原水凝胶已被用作药物和蛋白质输送支架[126,127]，以及肝脏[128]、皮肤[129]、血管[130]、小肠[131]、软骨[132]、声带[133]和脊髓[134]等器官的重建。

2.1.3 多肽类

2.1.3.1 透明质酸

透明质酸是一种线型双糖聚合物，由D-葡萄糖醛酸和N-乙酰-D-氨基葡萄糖通过交替的β-(1→4)-和β-(1→3)-之间的糖苷键连接（图2.7）[135,136]，存在于所有哺乳动物中，尤其是各种软结缔组织中，可以起到空间填充剂、润滑剂以及渗透缓冲剂的作用[137]。透明质酸可以在酰肼衍生物作用下共价交联成水凝胶[138,139]；并且其多糖结构上丰富的羧基和羟基为化学修饰提供了大量的活性位点[140]，如可与硫醇[141,142]、卤代乙酸酯[143]、二酰肼[138,144]、醛基[145,146]和酪胺基团[147]发生加成或缩合反应[148]。此外，透明质酸也可以被甲基丙烯酸酐或甲基丙烯酸缩水甘油酯改性，使其具有活性甲基丙烯酸基团，该基团可以引发自由基聚合[149-151]。由于透明质酸来源天然、非免疫原性、可生物降解和无黏附等特性[152-154]，透明质酸水凝胶可作为细胞治疗和组织工程中的支架，应用于细胞递送[155]、分子递送[156,157]、干细胞疗法[158,159]、软骨工程[156,160]、心脏修复[161]和瓣膜工程[162]。

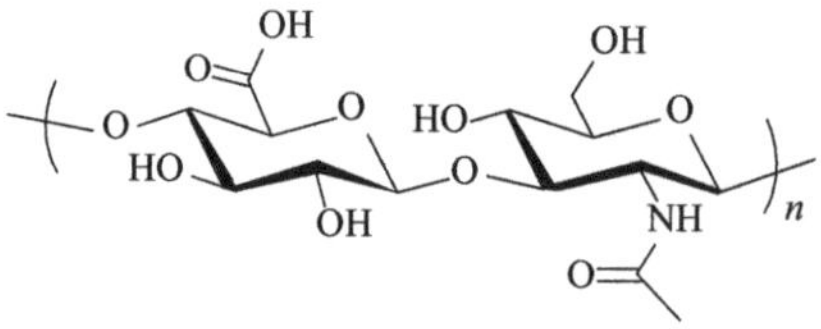

图 2.7 透明质酸的化学结构示意图

2.1.3.2 明胶

明胶是天然衍生的高分子，通过将胶原蛋白的三螺旋结构分解成单链分子而获得（图 2.8）；可分为 A 型和 B 型两类，分别由酸和碱处理胶原蛋白而得[163]。通常只需要将明胶水溶液的温度降低到特定温度以下就可以实现明胶的物理交联[164,165]。然而，物理交联的明胶不稳定，难以长期应用于生物医学领域；通常采用共价交联、化学接枝和物理掺杂等方式提高明胶水凝胶的力学性能和稳定性[167,179]。共价交联剂[166]如醛类（甲醛、戊二醛和甘油醛）[167,168]、聚环氧化物[169]和异氰酸酯[170]等，使其与明胶分子上的游离氨基（来自赖氨酸和羟赖氨酸）和 / 或游离羧酸（来自谷氨酸和天冬氨酸）反应桥联。与透明质酸类似，明胶主链也可以通过甲基丙烯酸酯进行改性，形成共价交联的明胶甲基丙烯酰水凝胶[171]。此外，有研究表明，通过化学接枝的方法将合成高分子偶联在明胶链上[172-174]，以及在明胶中掺杂可与其形成物理相互作用的高性能物质如碳纳米管[175]、氧化石墨烯[176]、无机纳米粒子、矿物质[177,178]等，均可增强明胶水凝胶的力学性能。明胶水凝胶易于凝胶化的特性以及出色的生物相容性，使其在药物输送[180]和组织工程[181,182]等生物医学中具有高应用价值。

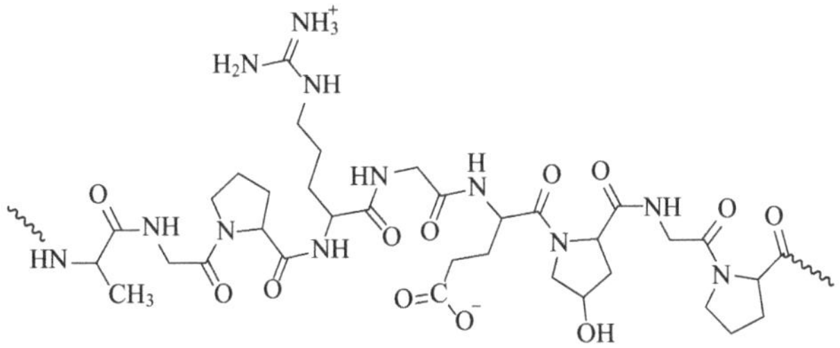

图 2.8　明胶的化学结构示意图

2.1.4　核酸类

脱氧核糖核酸（DNA）不仅可作为生物遗传的物质基础，又以其可编程性、功能多样性、生物相容性和生物可降解性等优点，在生物材料的构建方面表现出巨大的潜力[183]，其化学结构示意见图 2.9。DNA 水凝胶是一种主要由 DNA 参与形成的介于流体和固体之间的亲水三维网状聚合物材料，同时因其保留的 DNA 生物性能与自身骨架的力学性能的完美融合使其成为近年来最受关注的新兴功能高分子材料之一。通过固相合成和分子生物学技术，可以设计、合成具有某些特定序列的 DNA，实现可编程和可预测的方式精确制备一维、二维和三维 DNA 纳米结构[184-186]。通过设计，DNA 水凝胶可以呈现出不同的形状和尺寸，根据成分组成可分为单组分 DNA 水凝胶和多组分 DNA 水凝胶。水凝胶具有高负荷容量、优异的机械稳定性和黏弹性，可实现生物活性分子（如核苷酸、蛋白质、抗体和药物）的有效包封，并保留其生物活性[187-189]。通过向水凝胶序列中引入具备催化能力的脱氧核酶、有分子识别功能的适配体、能基因调节的 siRNA、有免疫激活能力的 CpG 寡核苷酸、pH 驱动的 C-G·C$^+$ 或 T-A·T 三链体等各类功能性核酸，可赋予水凝胶广泛的功能和特性[190-196]。由于 DNA 带有大量负电荷，其碱基具有芳香环结构特征，可与如丙烯酸酯、氨基、羧基和硫醇等反应，使其能够通过共价作用、静电吸引、氢键、范德华力等，与其他聚合物或纳米材料结合，从而拓展了 DNA 水凝胶的应用范围[197-199]。目前，基于各种功能碱基序列或通过结合不同的功能材料制备的单组分或多组分 DNA 水凝胶，已广泛用于生物医学、分子检测及环境保护的研究[200-202]。

图 2.9　DNA 的化学结构示意图

2.2 合成高分子

除了天然高分子外，合成高分子也已广泛用于水凝胶的设计和制造（图 2.10）。水凝胶的合成高分子网络可通过高分子骨架和交联剂的单体共聚，或合成高分子、高分子单体和交联剂的反应形成。

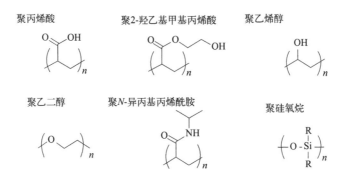

图 2.10 典型合成高分子的化学结构示意图

2.2.1 聚丙烯及其衍生物类

2.2.1.1 聚丙烯酸

聚丙烯酸（PAA）是通过丙烯酸单体的自由基聚合制备的线型聚合物。由于 PAA 的主链中含有大量的羧基，因此 PAA 可以通过共价或物理交联形成水凝胶。共价交联的 PAA 水凝胶通常由二/多乙烯基交联剂与丙烯酸单体共聚形成[203]。一方面，PAA 的羧基，不仅可以与不同的掺杂剂（如黏土[204]、氧化石墨烯[205] 和阳离子[206] 等）形成物理相互作用，还可以形成链间氢键，氢键的形成可为 PAA 水凝胶引入自修复性能[207]。另一方面，羧基可以与水分子结合，以促进 PAA 水凝胶对水的吸收[208]。由于羧基的化学特性，PAA 水凝胶的平衡溶胀比会受到水凝胶溶液的 pH 值以及离子强度的影响[209,210]。此外，PAA 水凝胶还可以结合其他的线型聚合物，如生物聚合物，以形成用于生物医学应用的各种黏合剂和水凝胶[211,212]。

2.2.1.2 聚 2- 羟乙基甲基丙烯酸

聚 2- 羟乙基甲基丙烯酸（PHEMA）水凝胶可通过 2- 羟乙基甲基丙烯酸（HEMA）单体与共价交联剂 [如丙二醇二甲基丙烯酸酯（TEGDMA）]、引发剂

[如焦亚硫酸钠（SMBS）和过硫酸铵（APS）] 的自由基聚合制备。HEMA 单体还可以与丙烯酸或丙烯酰胺单体共聚，以控制所得水凝胶的溶胀和力学性能[213]。纯 PHEMA 水凝胶具有抗细胞黏附性且不易在生理环境中降解，通过耦合不同生物活性基序，可改善其与细胞的相互作用和可降解性[214,215]。由于 PHEMA 水凝胶在生理环境中的优异光学透明性和力学稳定性，常常被应用于眼科领域，如隐形眼镜[216]和人工角膜[217]。

2.2.1.3 聚 N- 异丙基丙烯酰胺

丙烯酰胺及其衍生物，通过与交联剂的自由基共聚反应可制备水凝胶。聚 N- 异丙基丙烯酰胺（PNIPAM）水凝胶，作为性能优异的丙烯酰胺基水凝胶之一，受到了科研人员的广泛关注。当温度升高到临界温度以上时，未交联的线型 PNIPAM 会在水溶液中表现出从线圈到小球的相变[218,219]。PNIPAM 可以通过自由基聚合过程与双丙烯酰胺衍生物等交联剂实现共价交联，交联后的 PNIPAM 水凝胶会表现出可逆的热响应行为，临界温度约为 34℃[220]；高于该温度时，水凝胶的结构将发生坍塌并排出水[221,222]。PNIPAM 水凝胶的热响应行为通常很慢，但一些研究指出，通过在水凝胶形成过程中构建多孔结构可以提高 PNIPAM 水凝胶的响应速度[223,224]。这一类热响应 PNIPAM 水凝胶已被广泛应用于软致动器[225]、组织工程的可注射支架[226]以及用于按需分离细胞片的热响应基质[227,228]。

2.2.2 聚醇类

2.2.2.1 聚乙烯醇

聚乙烯醇（PVA）主要由聚醋酸乙烯酯部分水解得到[229]，可通过物理或共价交联形成稳定且有弹性的水凝胶[230,231]。物理交联的 PVA 水凝胶，通常可由反复冷冻和解冻 PVA 溶液制备[232]，具有弹性、坚韧、坚固和抗疲劳等特性[233-235]；化学交联的 PVA 水凝胶，还通过与双官能团交联剂（例如戊二醛、环氧氯丙烷、硼酸和二醛）进行共价交联而得[236,237]。此外，电子束和伽马（γ）辐射也可以交联 PVA，该方法有利于避免水凝胶中有共价交联剂残留[238]。通常，纯 PVA 水凝胶不黏附于细胞，但将几种寡肽序列缀合到 PVA 水凝胶的主链上，可增强它们与细胞之间的相互作用[239]。由于上述优异的特性，PVA 水凝胶已被广泛研究并用于生物医学应用[240,241]，如关节软骨置换和再生[242,243]。

2.2.2.2 聚乙二醇/聚环氧乙烷

聚乙二醇（PEG）通常由环氧乙烷的阴离子或阳离子聚合获得。当 PEG 的分子量超过 10×10^3 时，端基可以忽略不计，此时，也被称为聚环氧乙烷（PEO）[244]。

经过多年的研究，可以利用物理或化学交联方式将 PEG 聚合物交联成水凝胶。在化学交联方面，可以对 PEG 链端基进行化学反应。如用不饱和基团（丙烯酸酯或甲基丙烯酸酯末端）修饰 PEG 链的末端，然后将修饰后的高分子用作交联剂，通过光/紫外线诱导的自由基聚合与其他不饱和单体形成水凝胶[245,246]；还可以基于辐射诱导的自由基交联，通过电子束照射形成水凝胶[247]。此外，PEG 链的端基可通过各种"反应对"进行修饰，如 N- 羟基琥珀酰亚胺 /NH$_2$[248]、马来酰亚胺 /硫醇[249] 和乙炔 /叠氮化物[250]。由于这些功能性链端基序通常具有高反应效率和快速的反应动力学，因此通过这些"反应对"的偶联反应获得的水凝胶可以形成网络结构[251]。

与化学交联方法类似，PEG 链的末端可以用各种基序进行修饰以实现物理交联。例如，腺嘌呤和胸腺嘧啶的核酸碱基配对[252]、脲基-嘧啶酮单元[253] 或主客体分子[254] 均可以引入 PEG 分子的链端，以制备物理交联的 PEG 水凝胶。这些物理交联的 PEG 水凝胶能够表现出可切换、自修复、刺激响应特性以及较高的机械强度[255]。除了修饰和利用链端基团外，还可以使用 PEG 嵌段共聚物来制备物理交联的 PEG 水凝胶[256]。PEG-b-PPG（聚丙二醇）是最广泛使用的 PEG 衍生嵌段共聚物之一，由于 PPG 嵌段的疏水相互作用，可用于制备具有热响应特性的物理交联水凝胶[257]，并且可以调节平衡疏水性 PPG 嵌段和亲水性 PEG 嵌段两者的含量来优化所形成的水凝胶的相变行为。基于相同的凝胶化机制，PEG 嵌段共聚物与聚 DL- 乳酸（PDLLA）[258]、聚 DL- 乳酸 - 共 - 乙醇酸（PLGA）[259,260]、聚丙交酯（PLA）[261]、聚己内酯（PCL）[262] 和聚硫化丙烯（PPS）[263] 均可形成具有可注射或刺激响应特性的物理交联水凝胶。PEG 及其衍生物由于其无毒和非免疫原等特性而被广泛用于生物医学领域[264]。天然 PEG 水凝胶由于其生物惰性通常不会与细胞之间发生相互作用[265,266]，但利用各种生物活性结合物（生长因子[267] 和细胞黏附肽[268]）通过"迈克尔加成"或"点击化学"的手段，可对 PEG 水凝胶进行化学修饰[269,270]，使其形成所需的生物活性[266]。具有生物活性分子的 PEG 水凝胶可应用于药物/细胞输送[271,272] 和组织工程[273] 等。

2.2.3 其他类

有机硅水凝胶是指聚合物组分中包含了有机硅聚合物的水凝胶[274]。因为有机硅聚合物通常是疏水的[275]，所以为了形成有机硅水凝胶，会通过共混或共聚的方式将亲水性单体和/或聚合物引入有机硅基质中以提高有机硅水凝胶的亲水性[276-278]。例如，可以将 PHEMA 等亲水性聚合物直接混入有机硅聚合物基质中，形成亲水性互穿聚合物网络[279]。此外，N- 乙烯基吡咯烷酮（NVP）等亲水性单体也可以与硅大分子单体共聚形成亲水性有机硅水凝胶[280]。亲水性聚合物链段如 PEG[281] 能共聚到有机硅链段上，形成嵌段改性[282,283]或接枝改性[284]的亲水性有机硅水凝胶。由于这些亲水性有机硅水凝胶通常具有出色的透气性和良好的生物相容性，可用于制作隐形眼镜[276,285]，也可用作组织学工程材料[286,287]、药物输送载体[288,289]等。

参考文献

[1] Sabra W, Zeng A P, Deckwer W D. Bacterial alginate: physiology, product quality and process aspects. Applied microbiology and biotechnology, 2001, 56: 315-325.

[2] Lee K Y, Mooney D J. Alginate: properties and biomedical applications. Progress in polymer science, 2012, 37 (1): 106-126.

[3] Augst A D, Kong H J, Mooney D J. Alginate hydrogels as biomaterials. Macromolecular bioscience, 2006, 6 (8): 623-633.

[4] Donati I, Holtan S, Mørch Y A, et al. New hypothesis on the role of alternating sequences in calcium-alginate gels. Biomacromolecules, 2005, 6 (2): 1031-1040.

[5] Morris E R, Rees D A, Thom D. Characterisation of alginate composition and block-structure by circular dichroism. Carbohydrate Research, 1980, 81 (2): 305-314.

[6] Smidsrød O, Skja G. Alginate as immobilization matrix for cells. Trends in biotechnology, 1990, 8: 71-78.

[7] Pawar S N, Edgar K J. Alginate derivatization: A review of chemistry, properties and applications. Biomaterials, 2012, 33 (11): 3279-3305.

[8] Drury J L, Dennis R G, Mooney D J. The tensile properties of alginate hydrogels. Biomaterials, 2004, 25 (16): 3187-3199.

[9] Baer A E, Wang J Y, Kraus, V B, et al. Collagen gene expression and mechanical properties of intervertebral disc cell-alginate cultures. Journal of Orthopaedic Research, 2001, 19 (1): 2-10.

[10] Kang S W, Cha B H, Park H, et al. The effect of conjugating RGD into 3D alginate hydrogels on adipogenic differentiation of human adipose-derived stromal cells. Macromolecular bioscience, 2011, 11 (5): 673-679.

[11] Landa N, Miller L, Feinberg M S, et al. Effect of injectable alginate implant on cardiac remodeling and function after recent and old infarcts in rat. Circulation, 2008, 117 (11): 1388-1396.

[12] Selden C, Khalil M, Hodgson H. Three dimensional culture upregulates extracellular matrix protein expression in human liver cell lines-a step towards mimicking the liver in vivo? The International journal of artificial organs, 2000, 23 (11): 774-781.

[13] Marinho-Soriano E. Agar polysaccharides from Gracilaria species (Rhodophyta, Gracilariaceae). Journal of biotechnology, 2001, 89 (1): 81-84.

[14] Lahaye M, Rochas C. Chemical structure and physico-chemical properties of agar. Hydrobiologia, 1991, 221: 137-148.

[15] Normand V, Lootens D L, Amici E, et al. New insight into agarose gel mechanical properties. Biomacromolecules, 2000, 1 (4): 730-738.

[16] Uludag H, De Vos P, Tresco P A. Technology of mammalian cell encapsulation. Advanced drug delivery

reviews, 2000, 42 (1-2): 29-64.

[17] Dillon G P, Yu X, Sridharan A, et al. The influence of physical structure and charge on neurite extension in a 3D hydrogel scaffold. Journal of Biomaterials Science, Polymer Edition, 1998, 9 (10): 1049-1069.

[18] Luan N M, Iwata H. Xenotransplantation of islets enclosed in agarose microcapsule carrying soluble complement receptor 1. Biomaterials, 2012, 33 (32): 8075-8081.

[19] Kumachev A, Greener J, Tumarkin E, et al. High-throughput generation of hydrogel microbeads with varying elasticity for cell encapsulation. Biomaterials, 2011, 32 (6): 1477-1483.

[20] Clavé A, Potel J F, Servien E, et al. Third-generation autologous chondrocyte implantation versus mosaicplasty for knee cartilage injury: 2-year randomized trial. Journal of Orthopaedic Research, 2016, 34 (4): 658-665.

[21] Luo Y, Shoichet M S. A photolabile hydrogel for guided three-dimensional cell growth and migration. Nature materials, 2004, 3 (4): 249-253.

[22] Borkenhagen M, Clémence J F, Sigrist H, et al. Three-dimensional extracellular matrix engineering in the nervous system. Journal of biomedical materials research, 1998, 40 (3): 392-400.

[23] Bhattarai N, Gunn J, Zhang M. Chitosan-based hydrogels for controlled, localized drug delivery. Advanced drug delivery reviews, 2010, 62 (1): 83-99.

[24] Croisier F, Jérôme C. Chitosan-based biomaterials for tissue engineering. European polymer journal, 2013, 49 (4): 780-792.

[25] Sorlier P, Denuzière A, Viton C, et al. Relation between the degree of acetylation and the electrostatic properties of chitin and chitosan. Biomacromolecules, 2001, 2 (3): 765-772.

[26] Foster L J R, Ho S, Hook J, et al. Chitosan as a biomaterial: Influence of degree of deacetylation on its physiochemical, material and biological properties. PloS one, 2015, 10 (8): e0135153.

[27] Berger J, Reist M, Mayer J M, et al. Structure and interactions in covalently and ionically crosslinked chitosan hydrogels for biomedical applications. European journal of pharmaceutics and biopharmaceutics, 2004, 57 (1): 19-34.

[28] Boucard N, Viton C, Domard A. New aspects of the formation of physical hydrogels of chitosan in a hydroalcoholic medium. Biomacromolecules, 2005, 6 (6): 3227-3237.

[29] Brack H, Tirmizi S, Risen Jr W. A spectroscopic and viscometric study of the metal ion-induced gelation of the biopolymer chitosan. Polymer, 1997, 38 (10): 2351-2362.

[30] Dambies L, Vincent T, Domard A, et al. Preparation of chitosan gel beads by ionotropic molybdate gelation. Biomacromolecules, 2001, 2 (4): 1198-1205.

[31] Shu X, Zhu K. Controlled drug release properties of ionically cross-linked chitosan beads: the influence of anion structure. International Journal of pharmaceutics, 2002, 233 (1-2): 217-225.

[32] Shen E C, Wang C, Fu E, et al. Tetracycline release from tripolyphosphate-chitosan cross-linked sponge: a preliminary in vitro study. Journal of periodontal research, 2008, 43 (6): 642-648.

[33] Boddohi S, Moore N, Johnson P A, et al. Polysaccharide-based polyelectrolyte complex nanoparticles from chitosan, heparin, and hyaluronan. Biomacromolecules, 2009, 10 (6): 1402-1409.

[34] Denuziere A, Ferrier D, Damour O, et al. Chitosan-chondroitin sulfate and chitosan-hyaluronate polyelectrolyte complexes: biological properties. Biomaterials, 1998, 19 (14): 1275-1285.

[35] Jiang T, Zhang Z, Zhou Y, et al. Surface functionalization of titanium with chitosan/gelatin via electrophoretic deposition: characterization and cell behavior. Biomacromolecules, 2010, 11 (5): 1254-1260.

[36] Mao J S, Cui Y L, Wang X H, et al. A preliminary study on chitosan and gelatin polyelectrolyte complex cytocompatibility by cell cycle and apoptosis analysis. Biomaterials, 2004, 25 (18): 3973-3981.

[37] De Oliveira H, Fonseca J, Pereira M. Chitosan-poly (acrylic acid) polyelectrolyte complex membranes: preparation, characterization and permeability studies. Journal of Biomaterials Science, Polymer Edition, 2008, 19 (2): 143-160.

[38] Chenite A, Chaput C, Wang D, et al. Novel injectable neutral solutions of chitosan form biodegradable gels in situ. Biomaterials, 2000, 21 (21): 2155-2161.

[39] Zhang Y, Guan Y, Zhou S. Single component chitosan hydrogel microcapsule from a layer-by-layer approach. Biomacromolecules, 2005, 6 (4): 2365-2369.

[40] Milosavljević N B, Kljajević L M, Popović I G, et al. Chitosan, itaconic acid and poly (vinyl alcohol) hybrid polymer networks of high degree of swelling and good mechanical strength. Polymer International, 2010, 59 (5): 686-694.

[41] Singh A, Narvi S, Dutta P, et al. External stimuli response on a novel chitosan hydrogel crosslinked with formaldehyde. Bulletin of Materials Science, 2006, 29: 233-238.

[42] Liu R, Xu X, Zhuang X, et al. Solution blowing of chitosan/PVA hydrogel nanofiber mats. Carbohydrate polymers, 2014, 101: 1116-1121.

[43] Mi F L, Sung H W, Shyu S S. Synthesis and characterization of a novel chitosan-based network prepared using naturally occurring crosslinker. Journal of Polymer Science Part A: Polymer Chemistry, 2000, 38 (15): 2804-2814.

[44] Muzzarelli R A. Genipin-crosslinked chitosan hydrogels as biomedical and pharmaceutical aids. Carbohydrate Polymers, 2009, 77 (1): 1-9.

[45] Hoare T R, Kohane D S. Hydrogels in drug delivery: Progress and challenges. Polymer, 2008, 49 (8): 1993-2007.

[46] Hennink W E, van Nostrum C F. Novel crosslinking methods to design hydrogels. Advanced drug delivery reviews, 2012, 64: 223-236.

[47] Rickett T A, Amoozgar Z, Tuchek C A, et al. Rapidly photo-cross-linkable chitosan hydrogel for peripheral neurosurgeries. Biomacromolecules, 2011, 12 (1): 57-65.

[48] Cho I S, Cho M O, Li Z, et al. Synthesis and characterization of a new photo-crosslinkable glycol chitosan thermogel for biomedical applications. Carbohydrate polymers, 2016, 144: 59-67.

[49] Obara K, Ishihara M, Ishizuka T, et al. Photocrosslinkable chitosan hydrogel containing fibroblast growth factor-2 stimulates wound healing in healing-impaired db/db mice. Biomaterials, 2003, 24 (20): 3437-3444.

[50] Ahmadi F, Oveisi Z, Samani S M, et al. Chitosan based hydrogels: characteristics and pharmaceutical applications. Research in pharmaceutical sciences, 2015, 10 (1): 1.

[51] Cui Z K, Kim S, Baljon J J, et al. Microporous methacrylated glycol chitosan-montmorillonite nanocomposite hydrogel for bone tissue engineering. Nature communications, 2019, 10 (1): 3523.

[52] Kim I Y, Seo S J, Moon H S, et al. Chitosan and its derivatives for tissue engineering applications. Biotechnology advances, 2008, 26 (1): 1-21.

[53] Lee K Y, Ha W S, Park W H. Blood compatibility and biodegradability of partially N-acylated chitosan derivatives. Biomaterials, 1995, 16 (16): 1211-1216.

[54] Li J, Xu Z. Physical characterization of a chitosan-based hydrogel delivery system. Journal of pharmaceutical sciences, 2002, 91 (7): 1669-1677.

[55] Muzzarelli R, Zucchini C, Ilari P, et al. Osteoconductive properties of methylpyrrolidinone chitosan in an animal model. Biomaterials, 1993, 14 (12): 925-929.

[56] Elçin E, Elçin M, Pappas G. Neural tissue engineering: adrenal chromaffin cell attachment and viability on chitosan scaffolds. Neurological research, 1998, 20 (7): 648-654.

[57] Klemm D, Heublein B, Fink H P, et al. Cellulose: fascinating biopolymer and sustainable raw material. Angewandte chemie international edition, 2005, 44 (22): 3358-3393.

[58] Habibi Y, Lucia L A, Rojas O J. Cellulose nanocrystals: chemistry, self-assembly, and applications. Chemical reviews, 2010, 110 (6): 3479-3500.

[59] Moon R J, Martini A, Nairn J, et al. Cellulose nanomaterials review: structure, properties and nanocomposites. Chemical Society Reviews, 2011, 40 (7): 3941-3994.

[60] Ross P, Mayer R, Benziman M. Cellulose biosynthesis and function in bacteria. Microbiological reviews, 1991, 55 (1): 35-58.

[61] Czaja W K, Young D J, Kawecki M, et al. The future prospects of microbial cellulose in biomedical applications. Biomacromolecules, 2007, 8 (1): 1-12.

[62] Fink H P, Weigel P, Purz H, et al. Structure formation of regenerated cellulose materials from NMMO-solutions. Progress in Polymer Science, 2001, 26 (9): 1473-1524.

[63] Zhao H, Kwak J H, Wang Y, et al. Interactions between cellulose and N-methylmorpholine-N-oxide. Carbohydrate Polymers, 2007, 67 (1): 97-103.

[64] Zhang H, Wu J, Zhang J, et al. 1-Allyl-3-methylimidazolium chloride room temperature ionic liquid: a new and powerful nonderivatizing solvent for cellulose. Macromolecules, 2005, 38 (20): 8272-8277.

[65] Zhu S, Wu Y, Chen Q, et al. Dissolution of cellulose with ionic liquids and its application: a mini-review. Green Chemistry, 2006, 8 (4): 325-327.

[66] Cai J, Zhang L. Unique gelation behavior of cellulose in NaOH/urea aqueous solution. Biomacromolecules, 2006, 7 (1): 183-189.

[67] Liang S, Zhang L, Li Y, et al. Fabrication and properties of cellulose hydrated membrane with unique structure. Macromolecular Chemistry and Physics, 2007, 208 (6): 594-602.

[68] Li L, Thangamathesvaran P, Yue C, et al. Gel network structure of methylcellulose in water. Langmuir, 2001, 17 (26): 8062-8068.

[69] Xia X, Tang S, Lu X, et al. Formation and volume phase transition of hydroxypropyl cellulose microgels in salt solution. Macromolecules, 2003, 36 (10): 3695-3698.

[70] Weiss P, Gauthier O, Bouler J M, et al. Injectable bone substitute using a hydrophilic polymer. Bone, 1999, 25 (2): 67S-70S.

[71] Silva S M, Pinto F V, Antunes F E, et al. Aggregation and gelation in hydroxypropylmethyl cellulose

aqueous solutions. Journal of colloid and interface science, 2008, 327 (2): 333-340.

[72] Zheng W J, Gao J, Wei Z, et al. Facile fabrication of self-healing carboxymethyl cellulose hydrogels. European Polymer Journal, 2015, 72: 514-522.

[73] Kono H, Fujita S. Biodegradable superabsorbent hydrogels derived from cellulose by esterification crosslinking with 1, 2, 3, 4-butanetetracarboxylic dianhydride. Carbohydrate Polymers, 2012, 87 (4): 2582-2588.

[74] Yoshimura T, Matsuo K, Fujioka R. Novel biodegradable superabsorbent hydrogels derived from cotton cellulose and succinic anhydride: Synthesis and characterization. Journal of Applied Polymer Science, 2006, 99 (6): 3251-3256.

[75] Demitri C, Del Sole R, Scalera F, et al. Novel superabsorbent cellulose-based hydrogels crosslinked with citric acid. Journal of Applied Polymer Science, 2008, 110 (4): 2453-2460.

[76] Zhou J, Chang C, Zhang R, Zhang L. Hydrogels prepared from unsubstituted cellulose in NaOH/urea aqueous solution. Macromolecular Bioscience 2007, 7 (6): 804-809.

[77] Rodríguez R A, Alvarez-Lorenzo C, Concheiro A. Cationic cellulose hydrogels: kinetics of the cross-linking process and characterization as pH-/ion-sensitive drug delivery systems. Journal of Controlled Release, 2003, 86 (2-3): 253-265.

[78] Kabra B G, Gehrke S H, Spontak R J. Microporous, responsive hydroxypropyl cellulose gels. 1. Synthesis and microstructure. Macromolecules, 1998, 31 (7): 2166-2173.

[79] Fei B, Wach R A, et al. Hydrogel of biodegradable cellulose derivatives. I. Radiation-induced crosslinking of CMC. Journal of Applied Polymer Science, 2000, 78 (2): 278-283.

[80] Liu P, Peng J, Li J, et al. Radiation crosslinking of CMC-Na at low dose and its application as substitute for hydrogel. Radiation Physics and Chemistry, 2005, 72 (5): 635-638.

[81] Long D D, Luyen D V, Chitosan-carboxymethylcellulose hydrogels as supports for cell immobilization. Journal of Macromolecular Science, Part A: Pure and Applied Chemistry, 1996, 33 (12): 1875-1884.

[82] Hebeish A, Higazy A, El-Shafei A, et al. Synthesis of carboxymethyl cellulose (CMC) and starch-based hybrids and their applications in flocculation and sizing. Carbohydrate Polymers, 2010, 79 (1): 60-69.

[83] Işiklan N. Controlled release of insecticide carbaryl from sodium alginate, sodium alginate/gelatin, and sodium alginate/sodium carboxymethyl cellulose blend beads crosslinked with glutaraldehyde. Journal of applied polymer science, 2006, 99 (4): 1310-1319.

[84] Sannino A, Madaghiele M, Conversano F, et al. Cellulose derivative- hyaluronic acid-based microporous hydrogels cross-linked through divinyl sulfone (DVS) to modulate equilibrium sorption capacity and network stability. Biomacromolecules, 2004, 5 (1): 92-96.

[85] Liang S, Wu J, Tian H, et al. High-Strength Cellulose/Poly (ethylene glycol) Gels. ChemSusChem: Chemistry & Sustainability Energy & Materials, 2008, 1 (6): 558-563.

[86] Millon L, Wan W. The polyvinyl alcohol-bacterial cellulose system as a new nanocomposite for biomedical applications. Journal of Biomedical Materials Research Part B: Applied Biomaterials: An Official Journal of The Society for Biomaterials, The Japanese Society for Biomaterials, and The Australian Society for Biomaterials and the Korean Society for Biomaterials, 2006, 79 (2): 245-253.

[87] Williamson S L, Armentrout R S, Porter R S, et al. Microstructural examination of semi-interpenetrating networks of poly (N,N-dimethylacrylamide) with cellulose or chitin synthesized in lithium chloride/N, N-dimethylacetamide. Macromolecules, 1998, 31 (23): 8134-8141.

[88] Sannino A, Demitri C, Madaghiele M. Biodegradable cellulose-based hydrogels: design and applications. Materials, 2009, 2 (2): 353-373.

[89] Klemm D, Schumann D, Udhardt U, et al. Bacterial synthesized cellulose—artificial blood vessels for microsurgery. Progress in polymer science, 2001, 26 (9): 1561-1603.

[90] Chang C, Duan B, Cai J, et al. Superabsorbent hydrogels based on cellulose for smart swelling and controllable delivery. European polymer journal, 2010, 46 (1): 92-100.

[91] Vinatier C, Gauthier O, Fatimi A, et al. An injectable cellulose-based hydrogel for the transfer of autologous nasal chondrocytes in articular cartilage defects. Biotechnology and bioengineering, 2009, 102 (4): 1259-1267.

[92] Ye S H, Watanabe J, Iwasaki Y, et al. Antifouling blood purification membrane composed of cellulose acetate and phospholipid polymer. Biomaterials, 2003, 24 (23): 4143-4152.

[93] Hosseini H, Kokabi M, Mousavi S M. Conductive bacterial cellulose/multiwall carbon nanotubes nanocomposite aerogel as a potentially flexible lightweight strain sensor. Carbohydrate polymers, 2018, 201: 228-235.

[94] Zhou D, Zhang L, Zhou J, et al. Cellulose/chitin beads for adsorption of heavy metals in aqueous solution. Water research, 2004, 38 (11): 2643-2650.

[95] Noori A, Ashrafi S J, Vaez-Ghaemi R, et al. A review of fibrin and fibrin composites for bone tissue engineering. International journal of nanomedicine, 2017, 12: 4937.

[96] Perka C, Spitzer R S, Lindenhayn K, et al. Matrix-mixed culture: New methodology for chondrocyte culture and preparation of cartilage transplants. Journal of Biomedical Materials Research: An Official Journal of The Society for Biomaterials and The Japanese Society for Biomaterials, 2000, 49 (3): 305-311.

[97] Dare E V, Griffith M, Poitras P, et al. Genipin cross-linked fibrin hydrogels for in vitro human articular cartilage tissue-engineered regeneration. Cells Tissues Organs, 2009, 190 (6): 313-325.

[98] Lee C R, Grad S, Gorna K, et al. Fibrin-polyurethane composites for articular cartilage tissue engineering: a preliminary analysis. Tissue engineering, 2005, 11 (9-10): 1562-1573.

[99] Lieshout M V, Peters G, Rutten M, et al. A knitted, fibrin-covered polycaprolactone scaffold for tissue engineering of the aortic valve. Tissue engineering, 2006, 12 (3): 481-487.

[100] Weinand C, Gupta R, Huang A Y, et al. Comparison of hydrogels in the in vivo formation of tissue-engineered bone using mesenchymal stem cells and beta-tricalcium phosphate. Tissue Engineering, 2007, 13 (4): 757-765.

[101] Peled E, Boss J, Bejar J, et al. A novel poly (ethylene glycol) -fibrinogen hydrogel for tibial segmental defect repair in a rat model. Journal of biomedical materials research Part A, 2007, 80 (4): 874-884.

[102] Spotnitz W D, Prabhu R, Fibrin Sealant Tissue Adhesive-Review and Update. Journal of long-term effects of medical implants, 2005, 15 (3).

[103] Chrobak M O, Hansen K J, Gershlak J R, et al. Design of a fibrin microthread-based composite layer

for use in a cardiac patch. ACS biomaterials science & engineering, 2017, 3（7）: 1394-1403.

[104] Schense J C, Hubbell J A. Cross-linking exogenous bifunctional peptides into fibrin gels with factor XIIIa. Bioconjugate chemistry, 1999, 10（1）: 75-81.

[105] Han B, Schwab I R, Madsen T K, et al. A fibrin-based bioengineered ocular surface with human corneal epithelial stem cells. Cornea, 2002, 21（5）: 505-510.

[106] Meinhart J, Fussenegger M, Höbling W. Stabilization of fibrin-chondrocyte constructs for cartilage reconstruction. Annals of plastic surgery, 1999, 42（6）: 673-678.

[107] Johnson T S, Xu J W, Zaporojan V V, et al. Integrative repair of cartilage with articular and nonarticular chondrocytes. Tissue engineering, 2004, 10（9-10）: 1308-1315.

[108] Ye Q, Zünd G, Benedikt P, et al. Fibrin gel as a three dimensional matrix in cardiovascular tissue engineering. European Journal of Cardio-Thoracic Surgery, 2000, 17（5）: 587-591.

[109] Currie L J, Sharpe J R, Martin R. The use of fibrin glue in skin grafts and tissue-engineered skin replacements. Plast Reconstr Surg, 2001, 108: 1713-1726.

[110] Ahmed T A, Dare E V, Hincke M. Fibrin: a versatile scaffold for tissue engineering applications. Tissue Engineering Part B: Reviews, 2008, 14（2）: 199-215.

[111] Sorushanova A, Delgado L M, Wu Z, et al. The collagen suprafamily: from biosynthesis to advanced biomaterial development. Advanced materials, 2019, 31（1）: 1801651.

[112] Ferreira A M, Gentile P, Chiono V, et al. Collagen for bone tissue regeneration. Acta biomaterialia, 2012, 8（9）: 3191-3200.

[113] Shoulders M D, Raines R T. Collagen structure and stability. Annual review of biochemistry, 2009, 78: 929-958.

[114] Banwell E F, Abelardo E S, Adams D J, et al. Rational design and application of responsive α-helical peptide hydrogels. Nature materials, 2009, 8（7）: 596-600.

[115] Que R A, Arulmoli J, Da Silva N A, et al. Recombinant collagen scaffolds as substrates for human neural stem/progenitor cells. Journal of Biomedical Materials Research Part A, 2018, 106（5）: 1363-1372.

[116] Liu X, Zheng C, Luo X, et al. Recent advances of collagen-based biomaterials: Multi-hierarchical structure, modification and biomedical applications. Materials Science and Engineering: C, 2019, 99: 1509-1522.

[117] Schmitt F O, Hall C E, Jakus M A. Electron microscope investigations of the structure of collagen. Journal of Cellular and Comparative Physiology, 1942, 20（1）: 11-33.

[118] Glowacki J, Mizuno S. Collagen scaffolds for tissue engineering. Biopolymers: Original Research on Biomolecules, 2008, 89（5）: 338-344.

[119] Chan K L, Khankhel A H, Thompson R L, et al. Crosslinking of collagen scaffolds promotes blood and lymphatic vascular stability. Journal of biomedical materials research Part A, 2014, 102（9）: 3186-3195.

[120] Rault I, Frei V, Herbage D, et al. Evaluation of different chemical methods for cros-linking collagen gel, films and sponges. Journal of Materials Science: Materials in Medicine, 1996, 7（4）: 215-221.

[121] Chevallay B, Abdul-Malak N, Herbage D. Mouse fibroblasts in long-term culture within collagen

three-dimensional scaffolds: influence of crosslinking with diphenylphosphorylazide on matrix reorganization, growth, and biosynthetic and proteolytic activities. Journal of Biomedical Materials Research: An Official Journal of The Society for Biomaterials and The Japanese Society for Biomaterials, 2000, 49 (4): 448-459.

[122] Damink L O, Dijkstra P, Van Luyn M, et al. In vitro degradation of dermal sheep collagen crosslinked using a water-soluble carbodiimide. Biomaterials, 1996, 17 (7): 679-684.

[123] Couet F, Rajan N, Mantovani D. Macromolecular biomaterials for scaffold-based vascular tissue engineering. Macromolecular bioscience, 2007, 7 (5): 701-718.

[124] Marelli B, Achilli M, Alessandrino A, et al. Collagen-reinforced electrospun silk fibroin tubular construct as small calibre vascular graft. Macromolecular bioscience, 2012, 12 (11): 1566-1574.

[125] Pankajakshan D, Agrawal D K. Scaffolds in tissue engineering of blood vessels. Canadian journal of physiology and pharmacology, 2010, 88 (9): 855-873.

[126] Thacharodi D, Rao K P. Rate-controlling biopolymer membranes as transdermal delivery systems for nifedipine: development and in vitro evaluations. Biomaterials, 1996, 17 (13): 1307-1311.

[127] Lucas P A, Syftestad G T, Goldberg V M, et al. Ectopic induction of cartilage and bone by water-soluble proteins from bovine bone using a collagenous delivery vehicle. Journal of biomedical materials research, 1989, 23 (S13): 23-39.

[128] Kaufmann P, Heimrath S, Kim B, et al. Highly porous polymer matrices as a three-dimensional culture system for hepatocytes. Cell transplantation, 1997, 6 (5): 463-468.

[129] Helary C, Zarka M, Giraud-Guille M M, Fibroblasts within concentrated collagen hydrogels favour chronic skin wound healing. Journal of tissue engineering and regenerative medicine, 2012, 6 (3): 225-237.

[130] McGuigan A P, Sefton M V. The thrombogenicity of human umbilical vein endothelial cell seeded collagen modules. Biomaterials, 2008, 29 (16): 2453-2463.

[131] Voytik-Harbin S L, Brightman A O, Waisner B Z, et al. Small intestinal submucosa: A tissue-derived extracellular matrix that promotes tissue-specific growth and differentiation of cells in vitro. Tissue engineering, 1998, 4 (2): 157-174.

[132] Yuan T, Zhang L, Li K, et al. Collagen hydrogel as an immunomodulatory scaffold in cartilage tissue engineering. Journal of Biomedical Materials Research Part B: Applied Biomaterials, 2014, 102 (2): 337-344.

[133] Hahn M S, Teply B A, Stevens M M, et al. Collagen composite hydrogels for vocal fold lamina propria restoration. Biomaterials, 2006, 27 (7): 1104-1109.

[134] Iwata A, Browne K D, Pfister B J, et al. Long-term survival and outgrowth of mechanically engineered nervous tissue constructs implanted into spinal cord lesions. Tissue engineering, 2006, 12 (1): 101-110.

[135] Xu X, Jha A K, Harrington D A, et al. Hyaluronic acid-based hydrogels: from a natural polysaccharide to complex networks. Soft matter, 2012, 8 (12): 3280-3294.

[136] Burdick J A, Prestwich G D. Hyaluronic acid hydrogels for biomedical applications. Advanced materials, 2011, 23 (12): H41-H56.

[137] Laurent T C, Laurent U B, Fraser J R E. The structure and function of hyaluronan: an overview.

Immunology and cell biology, 1996, 74 (2): a1-a7.

[138] Vercruysse K P, Marecak D M, Marecek J F, et al. Synthesis and in vitro degradation of new polyvalent hydrazide cross-linked hydrogels of hyaluronic acid. Bioconjugate chemistry, 1997, 8 (5): 686-694.

[139] Pouyani T, Harbison G S, Prestwich G D, Novel hydrogels of hyaluronic acid: synthesis, surface morphology, and solid-state NMR. Journal of the American Chemical Society, 1994, 116 (17): 7515-7522.

[140] Kuo J, Prestwich G. Materials of biological origin-Materials analysis and implant uses, Comprehensive biomaterials. Ducheyne, P, 2010.

[141] Shu X Z, Liu Y, Palumbo F, et al. Disulfide-crosslinked hyaluronan-gelatin hydrogel films: a covalent mimic of the extracellular matrix for in vitro cell growth. Biomaterials, 2003, 24 (21): 3825-3834.

[142] Vanderhooft J L, Mann B K, Prestwich G D. Synthesis and characterization of novel thiol-reactive poly (ethylene glycol) cross-linkers for extracellular-matrix-mimetic biomaterials. Biomacromolecules, 2007, 8 (9): 2883-2889.

[143] Serban M A, Prestwich G D. Synthesis of hyaluronan haloacetates and biology of novel cross-linker-free synthetic extracellular matrix hydrogels. Biomacromolecules, 2007, 8 (9): 2821-2828.

[144] Pouyani T, Prestwich G D. Functionalized derivatives of hyaluronic acid oligosaccharides: drug carriers and novel biomaterials. Bioconjugate chemistry, 1994, 5 (4): 339-347.

[145] Jia X, Burdick J A, Kobler J, et al. Synthesis and characterization of in situ cross-linkable hyaluronic acid-based hydrogels with potential application for vocal fold regeneration. Macromolecules, 2004, 37 (9): 3239-3248.

[146] Jha A K, Hule R A, Jiao T, et al. Structural analysis and mechanical characterization of hyaluronic acid-based doubly cross-linked networks. Macromolecules, 2009, 42 (2): 537-546.

[147] Darr A, Calabro A. Synthesis and characterization of tyramine-based hyaluronan hydrogels. Journal of Materials Science: Materials in Medicine, 2009, 20: 33-44.

[148] Collins M N, Birkinshaw C. Hyaluronic acid based scaffolds for tissue engineering—A review. Carbohydrate polymers, 2013, 92 (2): 1262-1279.

[149] Burdick J A, Chung C, Jia X, et al. Controlled degradation and mechanical behavior of photopolymerized hyaluronic acid networks. Biomacromolecules, 2005, 6 (1): 386-391.

[150] Bian L, Guvendiren M, Mauck R L, et al. Hydrogels that mimic developmentally relevant matrix and N-cadherin interactions enhance MSC chondrogenesis. Proceedings of the National Academy of Sciences, 2013, 110 (25): 10117-10122.

[151] Bencherif S A, Srinivasan A, Horkay F, et al. Influence of the degree of methacrylation on hyaluronic acid hydrogels properties. Biomaterials, 2008, 29 (12): 1739-1749.

[152] Chen W J. Functions of hyaluronan in wound repair. Hyaluronan, 2002, 147-156.

[153] Kuo J. Practical aspects of hyaluronan based medical products. CRC Press, 2005.

[154] Prestwich G D. Engineering a clinically-useful matrix for cell therapy. Organogenesis, 2008, 4 (1): 42-47.

[155] Toh W S, Lee E H, Guo X M, et al. Cartilage repair using hyaluronan hydrogel-encapsulated human

embryonic stem cell-derived chondrogenic cells. Biomaterials, 2010, 31 (27): 6968-6980.

[156] Anseth K S, Metters A T, Bryant S J, et al. In situ forming degradable networks and their application in tissue engineering and drug delivery. Journal of controlled release, 2002, 78 (1-3): 199-209.

[157] Burdick J A, Ward M, Liang E, et al. Stimulation of neurite outgrowth by neurotrophins delivered from degradable hydrogels. Biomaterials, 2006, 27 (3): 452-459.

[158] Baksh D, Song L, Tuan R S. Adult mesenchymal stem cells: characterization, differentiation, and application in cell and gene therapy. Journal of cellular and molecular medicine, 2004, 8 (3): 301-316.

[159] Gerecht S, Burdick J A, Ferreira L S, et al. Hyaluronic acid hydrogel for controlled self-renewal and differentiation of human embryonic stem cells. Proceedings of the National Academy of Sciences, 2007, 104 (27): 11298-11303.

[160] Kim I L, Mauck R L, Burdick J A. Hydrogel design for cartilage tissue engineering: a case study with hyaluronic acid. Biomaterials, 2011, 32 (34): 8771-8782.

[161] Ifkovits J L, Tous E, Minakawa M, et al. Injectable hydrogel properties influence infarct expansion and extent of postinfarction left ventricular remodeling in an ovine model. Proceedings of the National Academy of Sciences, 2010, 107 (25): 11507-11512.

[162] Masters K S, Shah D N, Leinwand L A, et al. Crosslinked hyaluronan scaffolds as a biologically active carrier for valvular interstitial cells. Biomaterials, 2005, 26 (15): 2517-2525.

[163] Hudson C. Food hydrocolloids. Nishinari K and Doi E Plenum Press, New York, 1994.

[164] Segtnan V H, Isaksson T. Temperature, sample and time dependent structural characteristics of gelatine gels studied by near infrared spectroscopy. Food hydrocolloids, 2004, 18 (1): 1-11.

[165] Duconseille A, Astruc T, Quintana N, et al. Gelatin structure and composition linked to hard capsule dissolution: A review. Food hydrocolloids, 2015, 43: 360-376.

[166] Farris S, Song J, Huang Q. Alternative reaction mechanism for the cross-linking of gelatin with glutaraldehyde. Journal of agricultural and food chemistry, 2010, 58 (2): 998-1003.

[167] Kuijpers A, Engbers G, Feijen J, et al. Characterization of the network structure of carbodiimide cross-linked gelatin gels. Macromolecules, 1999, 32 (10): 3325-3333.

[168] Digenis G A, Gold T B, Shah V P. Cross-linking of gelatin capsules and its relevance to their in vitro-in vivo performance. Journal of pharmaceutical sciences, 1994, 83 (7): 915-921.

[169] Tu R, Shen S H, Lin D, et al. Fixation of bioprosthetic tissues with monofunctional and multifunctional polyepoxy compounds. Journal of biomedical materials research, 1994, 28 (6): 677-684.

[170] Hyndman C L, Groboillot A F, Poncelet D, et al. Microencapsulation of Lactococcus lactis within cross-linked gelatin membranes. Journal of Chemical Technology & Biotechnology, 1993, 56 (3): 259-263.

[171] Yue K, Trujillo-de Santiago G, Alvarez M M, et al. Synthesis, properties, and biomedical applications of gelatin methacryloyl (GelMA) hydrogels. Biomaterials, 2015, 73: 254-271.

[172] Morikawaand N, Matsuda T. Thermoresponsive artificial extracellular matrix: N-isopropylacrylamide-graft-copolymerized gelatin. Journal of Biomaterials Science, Polymer Edition, 2002, 13 (2): 167-183.

[173] Rizwan M, Yao Y, Gorbet M B, et al. One-pot covalent grafting of gelatin on poly (vinyl alcohol) hydrogel to enhance endothelialization and hemocompatibility for synthetic vascular graft applications. ACS applied bio materials, 2019, 3 (1): 693-703.

[174] Nichol J W, Koshy S T, Bae H, et al. Cell-laden microengineered gelatin methacrylate hydrogels. Biomaterials, 2010, 31 (21): 5536-5544.

[175] Shin S R, Bae H, Cha J M, et al. Carbon nanotube reinforced hybrid microgels as scaffold materials for cell encapsulation. ACS nano, 2011, 6 (1): 362-372.

[176] Paul A, Hasan A, Kindi H A, et al. Injectable graphene oxide/hydrogel-based angiogenic gene delivery system for vasculogenesis and cardiac repair. ACS nano, 2014, 8 (8): 8050-8062.

[177] Bakhsheshi-Rad H, Hadisi Z, Hamzah E, et al. Drug delivery and cytocompatibility of ciprofloxacin loaded gelatin nanofibers-coated Mg alloy. Materials Letters, 2017, 207: 179-182.

[178] Nagarajan S, Belaid H, Pochat-Bohatier C, et al. Design of boron nitride/gelatin electrospun nanofibers for bone tissue engineering. ACS applied materials & interfaces, 2017, 9 (39): 33695-33706.

[179] Choi Y S, Hong S R, Lee Y M, et al. Study on gelatin-containing artificial skin: I. Preparation and characteristics of novel gelatin-alginate sponge. Biomaterials, 1999, 20 (5): 409-417.

[180] Yamamoto M, Tabata Y, Ikada Y. Growth factor release from gelatin hydrogel for tissue engineering. Journal of bioactive and compatible polymers, 1999, 14 (6): 474-489.

[181] Sajkiewicz P, Kołbuk D. Electrospinning of gelatin for tissue engineering-molecular conformation as one of the overlooked problems. Journal of Biomaterials Science, Polymer Edition, 2014, 25 (18): 2009-2022.

[182] Panzavolta S, Gioffrè M, Focarete M L, et al. Electrospun gelatin nanofibers: optimization of genipin cross-linking to preserve fiber morphology after exposure to water. Acta biomaterialia, 2011, 7 (4): 1702-1709.

[183] Li F, Tang J, Geng J, et al. Polymeric DNA hydrogel: Design, synthesis and applications. Progress in Polymer Science, 2019, 98: 101163.

[184] Zhang Q, Lin S, Shi S, et al. Anti-inflammatory and antioxidative effects of tetrahedral DNA nanostructures via the modulation of macrophage responses. ACS applied materials & interfaces, 2018, 10 (4): 3421-3430.

[185] Tian T, Zhang T, Zhou T, et al. Synthesis of an ethyleneimine/tetrahedral DNA nanostructure complex and its potential application as a multi-functional delivery vehicle. Nanoscale, 2017, 9 (46): 18402-18412.

[186] Zhang H, Chao J, Pan D, et al. DNA origami-based shape IDs for single-molecule nanomechanical genotyping. Nature communications, 2017, 8 (1): 14738.

[187] Ji X, Wang Z, Niu S, et al. Non-template synthesis of porous carbon nanospheres coated with a DNA-cross-linked hydrogel for the simultaneous imaging of dual biomarkers in living cells. Chemical Communications, 2020, 56 (39): 5271-5274.

[188] Shao Y, Sun Z Y, Wang Y, et al. Designable immune therapeutical vaccine system based on DNA supramolecular hydrogels. ACS applied materials & interfaces, 2018, 10 (11): 9310-9314.

[189] He M, Nandu N, Uyar T B, et al. Small molecule-induced DNA hydrogel with encapsulation and

release properties. Chemical Communications, 2020, 56 (53): 7313-7316.

[190] Cheng E, Xing Y, Chen P, et al. A pH-triggered, fast-responding DNA hydrogel. Angewandte Chemie International Edition, 2009, 48 (41): 7660-7663.

[191] Hao L, Wang W, Shen X, et al. A fluorescent DNA hydrogel aptasensor based on the self-assembly of rolling circle amplification products for sensitive detection of ochratoxin A. Journal of agricultural and food chemistry, 2019, 68 (1): 369-375.

[192] Guo W, Lu C H, Orbach R, et al. pH-Stimulated DNA Hydrogels Exhibiting Shape-Memory Properties. Advanced Materials, 2015, 27 (1): 73-78.

[193] Alford A, Tucker B, Kozlovskaya V, et al. Encapsulation and ultrasound-triggered release of G-quadruplex DNA in multilayer hydrogel microcapsules. Polymers, 2018, 10 (12): 1342.

[194] Yan L, Zhu Z, Zou Y, et al. Target-responsive "sweet" hydrogel with glucometer readout for portable and quantitative detection of non-glucose targets. Journal of the American Chemical Society, 2013, 135 (10): 3748-3751.

[195] Simon A J, Walls-Smith L T, Freddi M J, et al. Simultaneous measurement of the dissolution kinetics of responsive DNA hydrogels at multiple length scales. ACS nano, 2017, 11 (1): 461-468.

[196] Lu S, Wang S, Zhao J, et al. A pH-controlled bidirectionally pure DNA hydrogel: reversible self-assembly and fluorescence monitoring. Chemical Communications, 2018, 54 (36): 4621-4624.

[197] Finke A, Schneider A K, Spreng A S, et al. Functionalized DNA Hydrogels Produced by Polymerase-Catalyzed Incorporation of Non-Natural Nucleotides as a Surface Coating for Cell Culture Applications. Advanced Healthcare Materials, 2019, 8 (9): 1900080.

[198] Li J, Mo L, Lu C H, et al. Functional nucleic acid-based hydrogels for bioanalytical and biomedical applications. Chemical Society Reviews, 2016, 45 (5): 1410-1431.

[199] Zhang Z, Han J, Pei Y, et al. Chaperone copolymer-assisted aptamer-patterned DNA hydrogels for triggering spatiotemporal release of protein. ACS Applied Bio Materials, 2018, 1 (4): 1206-1214.

[200] Zhang J, Guo Y, Pan G, et al. Injectable drug-conjugated DNA hydrogel for local chemotherapy to prevent tumor recurrence. ACS applied materials & interfaces, 2020, 12 (19): 21441-21449.

[201] Kahn J S, Ruiz R C, Sureka S, et al. DNA microgels as a platform for cell-free protein expression and display. Biomacromolecules, 2016, 17 (6): 2019-2026.

[202] Wang Y, Zhu Y, Hu Y, et al. How to construct DNA hydrogels for environmental applications: advanced water treatment and environmental analysis. Small, 2018, 14 (17): 1703305.

[203] Elliott J E, Macdonald M, Nie J, et al. Structure and swelling of poly (acrylic acid) hydrogels: effect of pH, ionic strength, and dilution on the crosslinked polymer structure. Polymer, 2004, 45 (5): 1503-1510.

[204] Zhao L, Huang J, Zhang Y, et al. Programmable and bidirectional bending of soft actuators based on Janus structure with sticky tough PAA-clay hydrogel. ACS Applied Materials & Interfaces, 2017, 9 (13): 11866-11873.

[205] Xie Z M, et al. Self-healable, super tough graphene oxide-poly (acrylic acid) nanocomposite hydrogels facilitated by dual cross-linking effects through dynamic ionic interactions. J Mater Chem B, 2015, 3 (19): 4001-4008.

[206] Zhao L, Huang J, Wang T, et al. Multiple Shape Memory, Self-Healable, and Supertough PAA-

GO-Fe^{3+} Hydrogel. Macromolecular Materials and Engineering, 2017, 302 (2): 1600359.

[207] Gulyuz U, Okay O. Self-healing poly (acrylic acid) hydrogels with shape memory behavior of high mechanical strength. Macromolecules, 2014, 47 (19): 6889-6899.

[208] Liu X, Steiger C, Lin S, et al. Ingestible hydrogel device. Nature communications, 2019, 10 (1): 493.

[209] Eichenbaum G M, Kiser P F, Dobrynin A V, et al. Investigation of the swelling response and loading of ionic microgels with drugs and proteins: The dependence on cross-link density. Macromolecules, 1999, 32 (15): 4867-4878.

[210] Scott R A, Peppas N A. Compositional effects on network structure of highly cross-linked copolymers of PEG-containing multiacrylates with acrylic acid. Macromolecules, 1999, 32 (19): 6139-6148.

[211] Peppas N A, Wright S L. Drug diffusion and binding in ionizable interpenetrating networks from poly (vinyl alcohol) and poly (acrylic acid). European Journal of Pharmaceutics and Biopharmaceutics, 1998, 46 (1): 15-29.

[212] Yuk H, Varela C E, Nabzdyk C S, et al. Dry double-sided tape for adhesion of wet tissues and devices. Nature, 2019, 575 (7781): 169-174.

[213] Sefton M, May M, Lahooti S, et al. Making microencapsulation work: conformal coating, immobilization gels and in vivo performance. Journal of Controlled Release, 2000, 65 (1-2): 173-186.

[214] Meyvis T, De Smedt S, Demeester J, et al. Influence of the degradation mechanism of hydrogels on their elastic and swelling properties during degradation. Macromolecules, 2000, 33 (13): 4717-4725.

[215] Canal T, Peppas N A. Correlation between mesh size and equilibrium degree of swelling of polymeric networks. Journal of biomedical materials research, 1989, 23 (10): 1183-1193.

[216] Kidane A, Szabocsik J M, Park K. Accelerated study on lysozyme deposition on poly (HEMA) contact lenses. Biomaterials, 1998, 19 (22): 2051-2055.

[217] Chirila T V. An overview of the development of artificial corneas with porous skirts and the use of PHEMA for such an application. Biomaterials, 2001, 22 (24): 3311-3317.

[218] Schild H G. Poly (*N*-isopropylacrylamide): experiment, theory and application. Progress in polymer science, 1992, 17 (2): 163-249.

[219] Tanaka F, Koga T, Winnik F M. Temperature-responsive polymers in mixed solvents: competitive hydrogen bonds cause cononsolvency. Physical review letters, 2008, 101 (2): 028302.

[220] Wu X S, Hoffman A S, Yager P. Synthesis and characterization of thermally reversible macroporous poly (*N*-isopropylacrylamide) hydrogels. Journal of Polymer Science Part A: Polymer Chemistry, 1992, 30 (10): 2121-2129.

[221] Xia L W, Xie R, Ju X J, et al. Nano-structured smart hydrogels with rapid response and high elasticity. Nature communications, 2013, 4 (1): 2226.

[222] Kumar A, Srivastava A, Galaev I Y, et al. Smart polymers: Physical forms and bioengineering applications. Progress in polymer science, 2007, 32 (10): 1205-1237.

[223] Serizawa T, Wakita K, Akashi M. Rapid deswelling of porous poly (*N*-isopropylacrylamide) hydrogels prepared by incorporation of silica particles. Macromolecules, 2002, 35 (1): 10-12.

[224] Ishida K, Uno T, Itoh T, et al. Synthesis and property of temperature-responsive hydrogel with movable cross-linking points. Macromolecules, 2012, 45 (15): 6136-6142.

[225] Breger J C, Yoon C, Xiao R, et al. Self-folding thermo-magnetically responsive soft microgrippers. ACS applied materials & interfaces, 2015, 7 (5): 3398-3405.

[226] Gutowska A, Jeong B, Jasionowski M. Injectable gels for tissue engineering. The Anatomical Record: An Official Publication of the American Association of Anatomists, 2001, 263 (4): 342-349.

[227] Brun-Graeppi A K A S, Richard C, Bessodes M, et al. Thermoresponsive surfaces for cell culture and enzyme-free cell detachment. Progress in Polymer Science, 2010, 35 (11): 1311-1324.

[228] Kim H, Witt H, Oswald T A, et al. Adhesion of epithelial cells to pnipam treated surfaces for temperature-controlled cell-sheet harvesting. ACS applied materials & interfaces, 2020, 12 (30): 33516-33529.

[229] Hassan C M, Peppas N A. Structure and morphology of freeze/thawed PVA hydrogels. Macromolecules, 2000, 33 (7): 2472-2479.

[230] Nuttelman C R, Mortisen D J, Henry S M, et al. Attachment of fibronectin to poly (vinyl alcohol) hydrogels promotes NIH3T3 cell adhesion, proliferation, and migration. Journal of Biomedical Materials Research: An Official Journal of The Society for Biomaterials, The Japanese Society for Biomaterials, and The Australian Society for Biomaterials and the Korean Society for Biomaterials, 2001, 57 (2): 217-223.

[231] Peppas N A, Merrill E W. Development of semicrystalline poly (vinyl alcohol) hydrogels for biomedical applications. Journal of biomedical materials research, 1977, 11 (3): 423-434.

[232] Stauffer S R, Peppast N A. Poly (vinyl alcohol) hydrogels prepared by freezing-thawing cyclic processing. Polymer, 1992, 33 (18): 3932-3936.

[233] Lin S, Liu J, Liu X, et al. Muscle-like fatigue-resistant hydrogels by mechanical training. Proceedings of the National Academy of Sciences, 2019, 116 (21): 10244-10249.

[234] Liu J, Lin S, Liu X, et al. Fatigue-resistant adhesion of hydrogels. Nature communications, 2020, 11 (1): 1071.

[235] Peppas N A, Stauffer S R. Reinforced uncrosslinked poly (vinyl alcohol) gels produced by cyclic freezing-thawing processes: a short review. Journal of Controlled Release, 1991, 16 (3): 305-310.

[236] Dai W, Barbari T. Hydrogel membranes with mesh size asymmetry based on the gradient crosslinking of poly (vinyl alcohol). Journal of Membrane Science, 1999, 156 (1): 67-79.

[237] Bo J. Study on PVA hydrogel crosslinked by epichlorohydrin. Journal of applied polymer science, 1992, 46 (5): 783-786.

[238] Yoshii F, Zhanshan Y, Isobe K, et al. Electron beam crosslinked PEO and PEO/PVA hydrogels for wound dressing. Radiation Physics and Chemistry, 1999, 55 (2): 133-138.

[239] Kobayashi H, Ikacia Y. Corneal cell adhesion and proliferation on hydrogel sheets bound with cell-adhesive proteins. Current eye research, 1991, 10 (10): 899-908.

[240] Bryant S, Nuttelman C, Anseth K. The effects of crosslinking density on cartilage formation in photocrosslinkable hydrogels. Biomedical sciences instrumentation, 1999, 35: 309-314.

[241] Shaheen S M, Yamaura K. Preparation of theophylline hydrogels of atactic poly (vinyl alcohol) / NaCl/H_2O system for drug delivery system. Journal of controlled release, 2002, 81 (3): 367-377.

[242] Gu Z Q, Xiao J M, Zhang X H. The development of artificial articular cartilage-PVA-hydrogel. Biomedical materials and engineering, 1998, 8 (2): 75-81.

[243] Oka M, Ushio K, Kumar P, et al. Development of artificial articular cartilage. Proceedings of the Institution of Mechanical Engineers, Part H: Journal of Engineering in Medicine, 2000, 214 (1): 59-68.

[244] Ostuni E, Chapman R G, Holmlin R E, et al. A survey of structure- property relationships of surfaces that resist the adsorption of protein. Langmuir, 2001, 17 (18): 5605-5620.

[245] Elbert D L, Hubbell J A. Conjugate addition reactions combined with free-radical cross-linking for the design of materials for tissue engineering. Biomacromolecules, 2001, 2 (2): 430-441.

[246] West J L, Hubbell J A. Photopolymerized hydrogel materials for drug delivery applications. Reactive Polymers, 1995, 25 (2-3): 139-147.

[247] Lopina S T, Wu G, Merrill E W, et al. Hepatocyte culture on carbohydrate-modified star polyethylene oxide hydrogels. Biomaterials, 1996, 17 (6): 559-569.

[248] Sakai T, Matsunaga T, Yamamoto Y, et al. Design and fabrication of a high-strength hydrogel with ideally homogeneous network structure from tetrahedron-like macromonomers. Macromolecules, 2008, 41 (14): 5379-5384.

[249] Ananda K, Nacharaju P, Smith P K, et al. Analysis of functionalization of methoxy-PEG as maleimide-PEG. Analytical biochemistry, 2008, 374 (2): 231-242.

[250] Malkoch M, Vestberg R, Gupta N, et al. Synthesis of well-defined hydrogel networks using Click chemistry. Chemical Communications, 2006 (26): 2774-2776.

[251] Kamata H, Akagi Y, Kayasuga-Kariya Y, et al. "Nonswellable" hydrogel without mechanical hysteresis. Science, 2014, 343 (6173): 873-875.

[252] Tan H, Xiao C, Sun J, et al. Biological self-assembly of injectable hydrogel as cell scaffold via specific nucleobase pairing. Chemical Communications, 2012, 48 (83): 10289-10291.

[253] Bastings M M, Koudstaal S, Kieltyka R E, et al. A fast pH-switchable and self-healing supramolecular hydrogel carrier for guided, local catheter injection in the infarcted myocardium. Advanced healthcare materials, 2014, 3 (1): 70-78.

[254] Chen X, Dong C, Wei K, et al. Supramolecular hydrogels cross-linked by preassembled host-guest PEG cross-linkers resist excessive, ultrafast, and non-resting cyclic compression. NPG Asia Materials, 2018, 10 (8): 788-799.

[255] Dong R, Pang Y, Su Y, et al. Supramolecular hydrogels: synthesis, properties and their biomedical applications. Biomaterials science, 2015, 3 (7): 937-954.

[256] Xue K, Liow S S, Karim A A, et al. A recent perspective on noncovalently formed polymeric hydrogels. The Chemical Record, 2018, 18 (10): 1517-1529.

[257] Kim H A, Lee H J, Hong J H, et al. α, ω-Diphenylalanine-end-capping of PEG-PPG-PEG polymers changes the micelle morphology and enhances stability of the thermogel. Biomacromolecules, 2017, 18 (7): 2214-2219.

[258] Shi K, Wang Y L, Qu Y, et al. Synthesis, characterization and application of reversible PDLLA-PEG-PDLLA copolymer thermogels in vitro and in vivo. Scientific reports, 2016, 6 (1): 19077.

[259] Sun J, Lei Y, Dai Z, et al. Sustained release of brimonidine from a new composite drug delivery

[259] system for treatment of glaucoma. ACS applied materials & interfaces, 2017, 9 (9): 7990-7999.

[260] Nagahama K, Oyama N, Ono K, et al. Nanocomposite injectable gels capable of self-replenishing regenerative extracellular microenvironments for in vivo tissue engineering. Biomaterials science, 2018, 6 (3): 550-561.

[261] Cui H, Shao J, Wang Y, et al. PLA-PEG-PLA and its electroactive tetraaniline copolymer as multi-interactive injectable hydrogels for tissue engineering. Biomacromolecules, 2013, 14 (6): 1904-1912.

[262] Gong C, Shi S, Dong P, et al. Synthesis and characterization of PEG-PCL-PEG thermosensitive hydrogel. International journal of pharmaceutics, 2009, 365 (1-2): 89-99.

[263] Zhang J, Tokatlian T, Zhong J, et al. Physically Associated Synthetic Hydrogels with Long-Term Covalent Stabilization for Cell Culture and Stem Cell Transplantation. Advanced Materials, 2011, 23 (43): 5098-5103.

[264] Zhao X, Milton Harris J. Novel degradable poly (ethylene glycol) hydrogels for controlled release of protein. Journal of pharmaceutical sciences, 1998, 87 (11): 1450-1458.

[265] Burdick J A, Anseth K S. Photoencapsulation of osteoblasts in injectable RGD-modified PEG hydrogels for bone tissue engineering. Biomaterials, 2002, 23 (22): 4315-4323.

[266] Mann B K, Tsai A T, Scott-Burden T, et al. Modification of surfaces with cell adhesion peptides alters extracellular matrix deposition. Biomaterials, 1999, 20 (23-24): 2281-2286.

[267] Kraehenbuehl T P, Zammaretti P, Van der Vlies A J, et al. Three-dimensional extracellular matrix-directed cardioprogenitor differentiation: systematic modulation of a synthetic cell-responsive PEG-hydrogel. Biomaterials, 2008, 29 (18): 2757-2766.

[268] Salinas C N, Anseth K S. Mixed mode thiol- acrylate photopolymerizations for the synthesis of PEG-peptide hydrogels. Macromolecules, 2008, 41 (16): 6019-6026.

[269] Metters A, Hubbell J. Network formation and degradation behavior of hydrogels formed by Michael-type addition reactions. Biomacromolecules, 2005, 6 (1): 290-301.

[270] Hiemstra C, van der Aa L J, Zhong Z, et al. Novel in situ forming, degradable dextran hydrogels by Michael addition chemistry: synthesis, rheology, and degradation. Macromolecules, 2007, 40 (4): 1165-1173.

[271] Stevens M M, Qanadilo H F, Langer R, et al. A rapid-curing alginate gel system: utility in periosteum-derived cartilage tissue engineering. Biomaterials, 2004, 25 (5): 887-894.

[272] García A J. PEG-maleimide hydrogels for protein and cell delivery in regenerative medicine. Annals of biomedical engineering, 2014, 42: 312-322.

[273] Zhu J. Bioactive modification of poly (ethylene glycol) hydrogels for tissue engineering. Biomaterials, 2010, 31 (17): 4639-4656.

[274] Tighe B J. A decade of silicone hydrogel development: surface properties, mechanical properties, and ocular compatibility. Eye & contact lens, 2013, 39 (1): 4-12.

[275] Eddington D T, Puccinelli J P, Beebe D J. Thermal aging and reduced hydrophobic recovery of polydimethylsiloxane. Sensors and Actuators B: Chemical, 2006, 114 (1): 170-172.

[276] Caló E, Khutoryanskiy V V. Biomedical applications of hydrogels: A review of patents and commercial products. European polymer journal, 2015, 65: 252-267.

[277] Peak C W, Wilker J J, Schmidt G. A review on tough and sticky hydrogels. Colloid and Polymer Science, 2013, 291, 2031-2047.

[278] Van Beek M, Weeks A, Jones L, et al. Immobilized hyaluronic acid containing model silicone hydrogels reduce protein adsorption. Journal of Biomaterials Science, Polymer Edition, 2008, 19 (11): 1425-1436.

[279] Tang Q, Yu J R, Chen L, et al. Poly (dimethyl siloxane) /poly (2-hydroxyethyl methacrylate) interpenetrating polymer network beads as potential capsules for biomedical use. Current Applied Physics, 2011, 11 (3): 945-950.

[280] Sugimoto H, Nishino G, Tsuzuki N, et al. Preparation of high oxygen permeable transparent hybrid copolymers with silicone macro-monomers. Colloid and Polymer Science, 2012, 290, 173-181.

[281] Zhang X, Wang L, Tao H, et al. The influences of poly (ethylene glycol) chain length on hydrophilicity, oxygen permeability, and mechanical properties of multicomponent silicone hydrogels. Colloid and Polymer Science, 2019, 297: 1233-1243.

[282] Lin G, Zhang X, Kumar S R, et al. Modification of polysiloxane networks for biocompatibility. Molecular Crystals and Liquid Crystals, 2010, 521 (1): 56-71.

[283] Yao M, Fang J. Hydrophilic PEO-PDMS for microfluidic applications. Journal of Micromechanics and Microengineering, 2012, 22 (2): 025012.

[284] Demming S, Lesche C, Schmolke H, et al. Characterization of long-term stability of hydrophilized PEG-grafted PDMS within different media for biotechnological and pharmaceutical applications. Physica status solidi (a), 2011, 208 (6): 1301-1307.

[285] Nicolson P C, Vogt J. Soft contact lens polymers: an evolution. Biomaterials, 2001, 22 (24): 3273-3283.

[286] Morales-Hurtado M, Zeng X, Gonzalez-Rodriguez P, et al. A new water absorbable mechanical Epidermal skin equivalent: The combination of hydrophobic PDMS and hydrophilic PVA hydrogel. Journal of the mechanical behavior of biomedical materials, 2015, 46: 305-317.

[287] Hamid Z A, Lim K. Evaluation of UV-crosslinked poly (ethylene glycol) diacrylate/poly (dimethylsiloxane) dimethacrylate hydrogel: properties for tissue engineering application. Procedia Chemistry, 2016, 19: 410-418.

[288] Xu J, Li X, Sun F. In vitro and in vivo evaluation of ketotifen fumarate-loaded silicone hydrogel contact lenses for ocular drug delivery. Drug delivery, 2011, 18 (2): 150-158.

[289] Wang J, Liu F, Wei J. Hydrophilic silicone hydrogels with interpenetrating network structure for extended delivery of ophthalmic drugs. Polymers for Advanced Technologies, 2012, 23 (9): 1258-1263.

第 3 章
水凝胶的合成方法

3.1 由温度引起的高分子链纠缠
3.2 分子自组装
3.3 离子凝胶化/静电相互作用
3.4 化学交联
3.5 小结
参考文献

水凝胶是由分散在水性介质中的高分子链通过不同的机制交联形成的，包括物理缠结、离子相互作用和化学交联（图 3.1）。大多数物理制备方法取决于高分子的固有性质，尽管这种依赖性限制了微调水凝胶属性的能力，但这类方法很容易实现且不需改变高分子链结构，并且在必要时很容易逆转。而化学交联的方法虽然可以实现更可控、更精确的交联过程，并能够以空间和时间动态定义的方式进行，但该交联过程难以逆转。

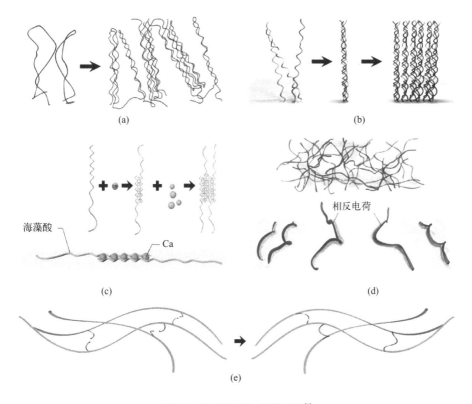

图 3.1 物理交联与化学交联[1]
（a）高分子链的热致缠结；（b）分子自组装；（c）离子凝胶化；
（d）静电相互作用；（e）化学交联

3.1 由温度引起的高分子链纠缠

许多天然聚合物，如海藻衍生的多糖和来自生物体的蛋白质，可以通过热驱动形成水凝胶。在胶凝过程中，高分子链会发生物理缠结以响应温度的变化。这种变化通常是由它们的溶解度改变以及具有物理刚性的填充高分子骨架的形成所引起的 [图 3.1（a）][2,3]。温度升高或降低均可能引起热凝胶化，对应的转变温度分别定义为下临界溶解温度（LCST）和上临界溶解温度（UCST）[4,5]。然而，胶凝机制因不同类型的高分子而异。通常可以表现出 UCST 的高分子包括明胶等天然高分子以及 PAA 等合成高分子，当温度降至低于其各自的 UCST 时，它们会凝胶化。与之相反，还有一些高分子会显示出 LCST 行为，如 PNIPAM 等合成高分子（图 3.2）在温度高于相应的 LCST 实现凝胶化。这一类热敏高分子的 LCST/UCST 可以通过它们的分子量、共聚物的比例、疏水/亲水链段的平衡来调节 [6,7]。

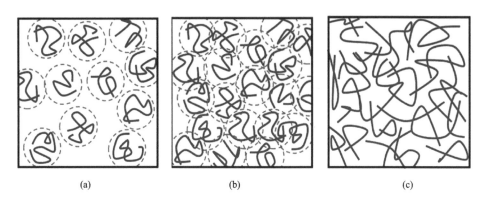

图 3.2　PNIPAM 的状态 [8]
（a）松散螺旋状态；（b）致密球态；（c）散装状态

3.2　分子自组装

非共价分子自组装也是用于构建水凝胶的一种常用策略，特别是对于基于蛋白质的水凝胶 [9]。弱的非共价键合机制，如氢键、范德华力、疏水相互作用等，使高分子折叠成具有明确结构和功能的骨架。值得注意的是，胶原蛋白的分层自组装过程 [图 3.1（b）]，依赖于富含脯氨酸或羟脯氨酸中氨基酸的规则排列 [10]。这些规则排列的分子可以促进称为原胶原的三股螺旋的形成，随后将原胶原亚基进一步包装成原纤维/纤维，最终形成胶原水凝胶 [9-11]。受这种机制的启发，仿生超分子水凝胶的设计可以遵循类似的分级自组装（图 3.3），例如胶原蛋白模拟肽 [11] 和基于肽两亲物和水凝胶剂的肽 [12,13]。

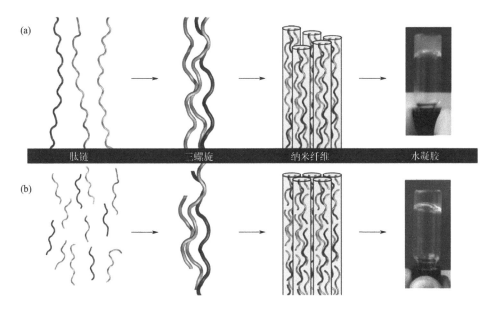

图 3.3 分子自组装

（a）I 型胶原蛋白组装示意图，其中肽链（以红色、蓝色和绿色显示）由 1000 个氨基酸组成，三螺旋长度为 100nm，平端纳米纤维（以灰色显示）通过三重螺旋的交错横向堆积组装；（b）胶原蛋白模拟肽的自组装方案，其中肽由 36 个氨基酸组成（以红色、蓝色和绿色显示），三股螺旋交错排列，长度为 10nm，纳米纤维（以灰色显示）由三螺旋伸长以及横向堆积引起

3.3 离子凝胶化/静电相互作用

自发的物理凝胶化还可能依赖于螯合或静电相互作用（图 3.4），例如，海藻酸盐基于螯合作用而形成水凝胶[14]。海藻酸盐中的 G 嵌段在遇到某些二价阳离子时会迅速凝胶化，如遇到 Ca^{2+} 或 Ba^{2+} 时，海藻酸盐会以"蛋盒"的形式凝胶化，其中成对的螺旋链堆积并包封内部的离子[图 3.1（c）][15]。天然高分子的复杂结构通常使其沿着骨架具有不同程度的静电荷。尽管许多天然高分子由于存在羧基而在中性 pH 值下带负电荷（如透明质酸和海藻酸盐），但对于一些氨基占主导地位的天然高分子在中性 pH 值下也可能呈现正电荷（例如明胶和壳聚糖）。相比之下，合成聚电解质的静电性能更容易控制，如聚 L-赖氨酸（PLL）/PAA 聚电解质对[16,17]，当这两种含有相反电荷的聚电解质的溶液混合时，其中的高分子会通过链缠结形成复合物，由于彼此之间相互屏蔽而变得不溶[图 3.1（d）][18]。

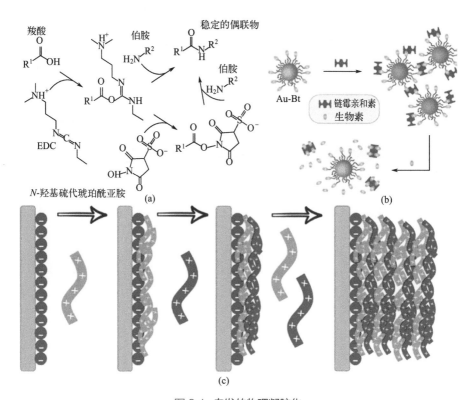

图 3.4 自发的物理凝胶化
（a）碳二亚胺反应的原理；（b）生物素 -（链霉菌）抗生物素蛋白结合示意图；（c）交替带电聚合物的逐层（LBL）组装[17]

3.4 化学交联

  一般而言，化学交联比物理方法能形成更稳定的水凝胶基质，因为它可以在凝胶化过程中显著提高成键的灵活性，实现高精度的时间和空间控制。水溶液中高分子主链或侧链上的化学活性结构/成分在适当的情况下可以形成共价键以获得水凝胶 [图 3.1（e）]。传统的机制包括缩合反应（如羟基/胺和羧酸之间的碳二亚胺化学）、自由基聚合、醛互补、高能辐射和酶促生物化学等[19]。在过去的十年中，"点击化学"经历了巨大的发展，作为一种被认为是快速、正交、高产的化学交联方法，广泛应用于细胞和生物活性剂。这类精确化学反应旨在解决与传统化学反应相关的挑战，如产率低、反应时间长和反应条件极端等[20-22]。此外，研究人员还设计出了多种具有生物相容性的"点击化学"反应（图 3.5），用于在基质内存在细胞的情况下直接通过生物偶联形成固态水凝胶，如硫醇 - 乙烯基砜和硫醇-马来酰亚胺"迈克尔加成"反应、叠氮-炔和叠氮-炔环加成反应、噻吩光电偶联反应等[22]。

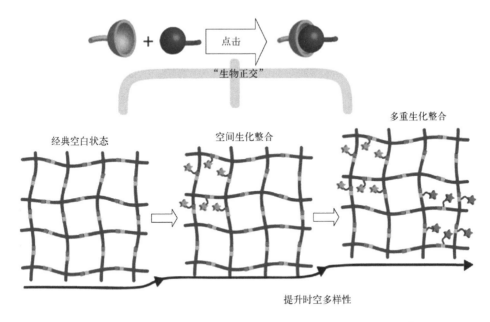

图 3.5　生物偶联或水凝胶交联中常用的"点击化学"反应示意图[22]

3.5　小结

上述每一种凝胶化方法都有其优点和局限性，笔者相信该领域会向多种机制结合，促进水凝胶结构与性能优化的方向发展[23-26]。这些基于多种机制构造的水凝胶通常会表现出优异的物理化学性质，如显著增强的力学性能、可注射性、自我修复和进行动态调节的可能性，这些特性将在后续章节中进一步讨论。

参考文献

[1] Zhang Y S, Khademhosseini A. Advances in engineering hydrogels. Science, 2017, 356（6337）: eaaf3627.

[2] Djabourov M, Leblond J, Papon P. Gelation of aqueous gelatin solutions Ⅰ. Structural investigation. Journal de physique, 1988, 49（2）: 319-332.

[3] Djabourov M, Leblond J, Papon P. Gelation of aqueous gelatin solutions Ⅱ. Rheology of the sol-gel transition. Journal de Physique, 1988, 49（2）: 333-343.

[4] Gasperini L, Mano J F, Reis R L. Natural polymers for the microencapsulation of cells. Journal of the royal society Interface, 2014, 11 (100): 20140817.

[5] Ward M A, Georgiou T K. Thermoresponsive polymers for biomedical applications. Polymers, 2011, 3 (3): 1215-1242.

[6] Lutz J F, Akdemir Ö, Hoth A. Point by point comparison of two thermosensitive polymers exhibiting a similar LCST: is the age of poly (NIPAM) over? Journal of the American Chemical Society, 2006, 128 (40): 13046-13047.

[7] Laschewsky A, Rekaï E D, Wischerhoff E. Tailoring of stimuli-responsive water soluble acrylamide and methacrylamide polymers. Macromolecular Chemistry and Physics, 2001, 202 (2): 276-286.

[8] Kong D C, Yang M H, Zhang X S, et al. Control of polymer properties by entanglement: a review. Macromolecular Materials and Engineering, 2021, 306 (12): 2100536.

[9] Zhang S. Fabrication of novel biomaterials through molecular self-assembly. Nature biotechnology, 2003, 21 (10): 1171-1178.

[10] Rich A, Crick F. The structure of collagen. Nature, 1955, 176: 915-916.

[11] O'leary L E, Fallas J A, Bakota E L, et al. Multi-hierarchical self-assembly of a collagen mimetic peptide from triple helix to nanofibre and hydrogel. Nature chemistry, 2011, 3 (10): 821-828.

[12] Hartgerink J D, Beniash E, Stupp S I. Self-assembly and mineralization of peptide-amphiphile nanofibers. Science, 2001, 294 (5547): 1684-1688.

[13] Yang Z, Xu B. Supramolecular hydrogels based on biofunctional nanofibers of self-assembled small molecules. Journal of Materials Chemistry, 2007, 17 (23): 2385-2393.

[14] Rowley J A, Madlambayan G, Mooney D J. Alginate hydrogels as synthetic extracellular matrix materials. Biomaterials, 1999, 20 (1): 45-53.

[15] Braccini I, Pérez S. Molecular basis of Ca^{2+}-induced gelation in alginates and pectins: the egg-box model revisited. Biomacromolecules, 2001, 2 (4): 1089-1096.

[16] Morton S W, Herlihy K P, Shopsowitz K E, et al. Scalable manufacture of built-to-order nanomedicine: spray-assisted layer-by-layer functionalization of PRINT nanoparticles. Advanced Materials, 2013, 25 (34): 4707-4713.

[17] Leijten J, Rouwkema J, Zhang Y S, et al. Advancing tissue engineering: a tale of nano-, micro-, and macroscale integration. Small, 2016, 12 (16): 2130-2145.

[18] Bekturov E A, Bimendina L A, Complexes of water-soluble polymers. Journal of Macromolecular Science, Part C: Polymer Reviews, 1997, 37 (3): 501-518.

[19] Hennink W E, van Nostrum C F, Novel crosslinking methods to design hydrogels. Advanced drug delivery reviews, 2012, 64, 223-236.

[20] DeForest C A, Polizzotti B D, Anseth K S, Sequential click reactions for synthesizing and patterning three-dimensional cell microenvironments. Nature materials, 2009, 8 (8): 659-664.

[21] DeForest C A, Anseth K S, Cytocompatible click-based hydrogels with dynamically tunable properties through orthogonal photoconjugation and photocleavage reactions. Nature chemistry, 2011, 3 (12): 925-931.

[22] Azagarsamy M A, Anseth K S, Bioorthogonal click chemistry: an indispensable tool to create multifaceted cell culture scaffolds. ACS Publications, 2013.

[23] Li J, Illeperuma W R, Suo Z, et al. Hybrid hydrogels with extremely high stiffness and toughness. ACS Macro Letters, 2014, 3 (6): 520-523.

[24] Annabi N, Shin S R, Tamayol A, et al. Highly elastic and conductive human-based protein hybrid hydrogels. Advanced Materials, 2016, 28 (1): 40-49.

[25] Yuk H, Zhang T, Parada G A, et al. Skin-inspired hydrogel-elastomer hybrids with robust interfaces and functional microstructures. Nature communications, 2016, 7 (1): 12028.

[26] Gaharwar A K, Peppas N A, Khademhosseini A. Nanocomposite hydrogels for biomedical applications. Biotechnology and bioengineering, 2014, 111 (3): 441-453.

第4章
水凝胶内部的主要交联类型

4.1 永久共价交联
4.2 强物理交联
4.3 弱物理交联
4.4 动态共价交联
参考文献

在本章中，我们将讨论水凝胶中常见的交联类型——永久共价交联网络和 UPN 相互作用网络。其中，UPN 相互作用被定义为不同于传统聚合物网络（即永久共价交联）的聚合物间和聚合物内相互作用。根据 UPN 相互作用的性质，可大致分为三类：强物理交联、弱物理交联和动态共价交联。

4.1 永久共价交联

水凝胶中常见的永久共价交联键包含碳-碳键、碳-氮键、碳-氧键、碳-硫键和硅-氧键等（图 4.1），其能量范围通常为 220～570kJ/mol（图 4.2）[1-3]。

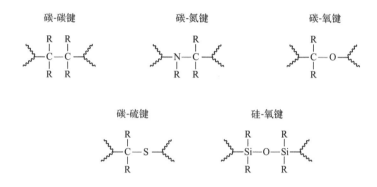

图 4.1 水凝胶中常见的永久共价交联键（R 代表有机基取代基或氢）

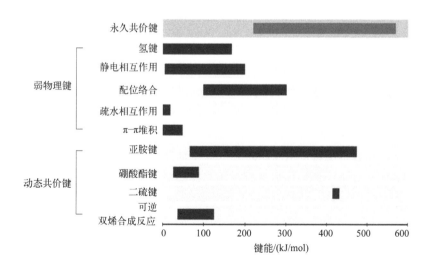

图 4.2 永久共价交联、弱物理交联和动态共价交联的键能

4.1.1 碳 - 碳键

碳 - 碳键的键能为 300～450kJ/mol[1-3]。通过碳 - 碳键共价交联的水凝胶通常由单体和二/多乙烯基交联剂通过自由基共聚形成。交联剂可以是具有两个双键的小分子，如 N,N'- 亚甲基双丙烯酰胺（MBA）或具有多个丙烯酸酯基的大分子[4,5]。这些交联剂适用于各种引发和聚合体系[4,6,7]。将光自由基引发剂、单体、二/多乙烯基交联剂共同加入预聚合溶液中[8-10]，一旦引发剂被紫外线照射，就会产生自由基以引发单体和交联剂上双键的聚合[11,12]。因此，该方法可以原位形成具有图案化结构或生物功能的水凝胶[9,13]。此外，乙烯基单体和交联剂的聚合也可以在过二硫酸盐和 $N,N,N'N'$- 四亚甲基二胺（TEMED）组成的体系中进行。其中，TEMED 可以加速过二硫酸盐的分解，并生成大量自由基从而引发聚合反应，由此，可在室温下快速形成各种水凝胶[14]。

水凝胶的碳 - 碳交联也可以通过高能辐射（如伽马射线、电子束等）形成。与紫外线类似，除了能交联如带有乙烯基或丙烯酸酯基的单体和交联剂[15,16]的聚合不饱和化合物外，高能辐射由于可以引起聚合物链均裂断裂并产生自由基，还可交联没有不饱和键的聚合物[17]。需要注意的是，交联过程中要尽量使用无水溶剂，主要是因为溶剂中水分子的辐射分解会产生羟基自由基，并攻击聚合物链从而形成大分子自由基，然后进一步通过这些自由基的重组和终止形成由碳 - 碳键交联的共价聚合物网络[18]。

4.1.2 碳 - 氮键

碳 - 氮键的键能为 300～430kJ/mol[1-3]。由碳 - 氮键共价交联的水凝胶通常由互补基团的高效化学反应形成。目前常用的形成碳 - 氮键的方式有三种。第一种为胺与羧酸及其衍生物之间的缩合反应形成的酰胺键[19]。通常将 N- 羟基琥珀酰亚胺（NHS）和 N,N-(3-二甲氨基丙基)-N- 乙基碳二亚胺（EDC）作为胺与羧酸的缩合反应的促进剂[20]，可有效抑制副反应，并控制所形成的水凝胶的交联密度[21]。第二种为胺与亲电子试剂（如己二酸二酰肼和二异氰酸酯交联剂）的加成反应形成[4,22,23]。该反应的反应效率高，所得水凝胶的力学性能可通过调节聚合物和交联剂的浓度和比例来控制，已被广泛用于交联天然大分子。第三种为叠氮化物与炔烃的环加成反应。这是一种典型的"点击反应"，可将炔烃和叠氮化合物形成三唑结构，具有反应效率高、无副反应等优点[24]。此外，叠氮化物 - 炔烃环之间的加成反应可以在没有金属催化的情况下进行，这扩大了叠氮化物 - 炔烃环加成反应在制备生物相容性水凝胶方面的适用性[25]。

4.1.3 碳-氧键

碳-氧键的键能为 280～370kJ/mol[1-3]。最常见的碳-氧键是由羟基与羧酸及其衍生物反应形成的酯键[26]。这种酯交联很容易水解，使得水凝胶在环境温度和生理条件下实现降解。除酯交联外，碳-氧键也存在于醚基和氨基甲酸酯基中，由于聚合物上的侧基（如多糖和 PVA 上的羟基）会和交联剂（如戊二醛[27,28]、二乙烯基砜[29]、二溴化物[30]、二异氰酸酯[31] 等）发生反应，从而形成化学交联的碳-氧键。

4.1.4 碳-硫键

碳-硫键的键能为 220～310kJ/mol[3]。水凝胶通过碳-硫键实现的共价交联主要由硫醇的"点击反应"形成[32,33]。硫原子的高电子密度使硫醇易于以自由基的形式与多种官能团发生反应[34,35]。硫醇基团可以很容易地转化为亲核硫醇盐或亲电硫基自由基，然后进行基于亲核加成或自由基链过程实现的硫醇"点击反应"[36]。具体来说，对于自由基硫醇"点击反应"，硫醇基团可以通过高温和/或紫外线激活而产生自由基，进而引发自由基介导的硫醇-烯或硫醇-炔反应[37]。对于由强碱引发的亲核硫醇"点击反应"，硫醇基团可以很容易地通过迈克尔加成与缺电子的烯功能化合物反应，通过羰基加成与异氰酸酯衍生物反应，通过 S_N2 亲核取代与卤化物反应，并通过 S_N2 开环与环氧序反应[38-41]。硫醇"点击反应"通常具有高效率和高转化率的特点，即使在有水、离子、氧气存在的情况下，也没有任何副产物，已被广泛用于制备各种生物医学应用的水凝胶[42,43]。

4.1.5 硅-氧键

硅-氧键的键能为 420～570kJ/mol[1-3]。硅-氧键主要用于硅基水凝胶的形成[44-46]，可以有效提高硅基水凝胶的力学性能[47]，被广泛用于键合水凝胶和各种具有改性表面（如盐化表面）的工程材料[48]。

4.2 强物理交联

除了 4.1 节中讨论的永久共价交联之外，各种类型的强物理键同样可以在聚合物网络中充当有效的永久交联，其键能与永久共价交联的键能相似。强物理交联的典型例子包括晶畴、玻璃状结节和螺旋关联。

4.2.1 晶畴

合成和天然高分子的特定子集可以在适当条件下形成晶畴。尺寸从纳米到微米的晶畴可以作为强物理交联连接多个无定形聚合物链（图 4.3）。作为合成高分子的一个例子，PVA 可以通过反复的冻融循环或在高于其玻璃化转变温度的温度下退火来形成晶畴[8,49,50]。PVA 晶畴的形成主要是由于 PVA 链上羟基之间的氢键相互作用[51]。天然高分子，可以通过用强酸或强碱溶液处理几丁质和壳聚糖来克服它们的链间静电排斥，从而形成具有晶畴交联的无定形链半结晶高分子网络[52,53]；由于葡萄糖单元之间的强相互作用，纤维素也可以形成高度结晶的纳米纤维[54]，这些纤维素纳米纤维可以通过碱处理实现进一步聚集并形成稳定的网络[55,56]。值得注意的是，尽管大多数晶畴在室温和体温下都是稳定的，但当上述半结晶高分子网络的加热温度高于其熔化温度时，其网络中的晶畴会被破坏。

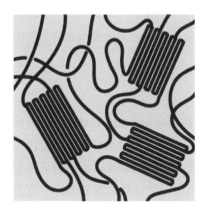

图 4.3　基于晶畴实现强物理交联的示意图

由于晶畴通常互连多个高分子链，因此它们通常充当高分子网络中的高功能交联键。此外，将高分子链拉出晶畴所需的能量远高于使同一高分子链断裂所需的能量[57]。因此，晶畴也可以充当高分子网络中的固有高能相。晶畴的这些属性赋予了含有晶畴的水凝胶高韧、高强、高弹、抗疲劳等极端的力学性能，这将在第 6 章中讨论。

4.2.2 玻璃状结节

当温度低于其玻璃化转变温度时，无定形聚合物通过可逆液态玻璃化转变形成玻璃状结节[58]。为了利用玻璃状结节作为强物理交联，在构建水凝胶时通常使用至少包含一个具有高玻璃化转变温度链段的嵌段共聚物。随着温度降至室温，具有高玻璃化转变温度的链段会形成玻璃状结节，这可以有效地交联相邻的无定形聚合物链（图 4.4）[59]。例如，聚苯乙烯 -b- 聚 N- 异丙基丙烯酰胺 -b-

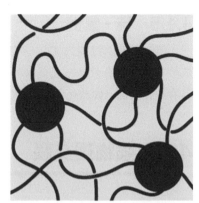

图 4.4　基于玻璃状结节实现强物理交联的示意图

聚苯乙烯共聚物中的聚苯乙烯链段可以在室温下形成玻璃状结节，将嵌段共聚物链交联成聚合物网络[60]。聚甲基丙烯酸甲酯的玻璃化转变温度约为115℃[61]；因此，聚甲基丙烯酸甲酯-b-聚丙烯酸正丁酯共聚物中的聚甲基丙烯酸甲酯链段可以在室温下形成玻璃状结节以交联聚合物网络[62]。与晶畴类似，玻璃状结节也可以作为聚合物网络中的高功能交联键和固有高能相，使相应的水凝胶具有较好的力学性能，这将在第6章中讨论。

4.2.3 螺旋关联

许多天然高分子，由于其精确控制的结构，可以组装成纳米级的螺旋纤维（或原纤维），并进一步通过聚集或缠结形成交联网络（图4.5）[23,63-65]。Ⅰ型胶原三螺旋结构是由三个肽链自组装形成的。这些胶原蛋白三螺旋可以堆积在一起形成胶原蛋白纳米纤维，并自组装成相互连接的水凝胶网络[66,67]。而线型琼脂糖链在高温水溶液中是无序的分子，但当温度降低到室温或体温时，就可以形成双螺旋链[68]或单螺旋链[69]。这些螺旋链可以通

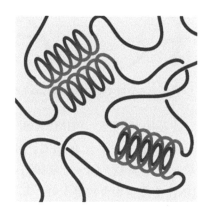

图4.5 基于螺旋关联实现强物理交联的示意图

过氢键的方式结合形成琼脂糖纤维，并进一步缠结成相互连接的水凝胶网络[70]。

4.3 弱物理交联

除了强物理交联外，高分子网络中还有一些相对较弱、短暂且可逆的其他物理交联，其能量通常低于强物理交联和共价交联的能量（图4.2），主要包括氢键、静电相互作用、金属配位、主客体相互作用、疏水缔合、π-π 堆积等。

4.3.1 氢键

单个氢键的能量范围为 0.8 ~ 167kJ/mol（图4.2）[71,72]。许多天然高分子可以通过分子间氢键形成水凝胶（图4.6）。例如，明胶可以形成由氢键交联的螺旋状高分子网络[73]；某些多糖（如琼脂糖、直链淀粉、支链淀粉和角叉菜胶）也可以在溶液中形成螺旋结构，并通过氢键交联成水凝胶[74]。此外，大量合成高分子也能够通过氢键形成物理水凝胶。例如，PVA 水凝胶可以通过 PVA 溶液的反复冷冻

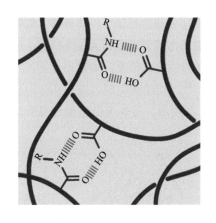

图 4.6 基于氢键实现弱物理交联的示意图

和解冻在聚合物链之间形成氢键而获得[75]。聚甲基丙烯酸（PMA）或 PAA 可通过 PEG 的氧基团与 PMA 或 PAA 的羧基之间的氢键形成复合物[76,77]。

尽管天然高分子和合成高分子中存在丰富的氢键基团（—OH、—NH、—C=O、—O—），但由于大量水分子的存在，水凝胶中的氢键相互作用通常会被削弱。为了实现有效的氢键交联，通常会将具有多个自互补氢键基团的疏水嵌段引入聚合物中[78-80]。例如，当用三胺嗪或二氨基三嗪基团对 PEG、PHMEA 和 PNIPAM 功能改性后，可以在交联网络中形成三重氢键[78,79,81,82]。在 PEG、PHMEA、PNIPAM、PAA 和 PDMAA 链上引入脲基吡啶亚胺酮（UPy）基团可在交联网络中形成四重氢键[78-80]。此外，当将 DNA 连接到聚合物链上时，互补的 DNA 碱基对（A-T、C-G）也可以作为氢键基序[83]。

4.3.2 静电相互作用

静电相互作用的能量范围为 5～200kJ/mol（图 4.2）[84]。聚电解质，是具有固定电荷的天然高分子和合成高分子，可通过静电相互作用实现物理交联（图 4.7）[85-88]。海藻酸盐作为阴离子聚电解质，可与多种二价阳离子如 Ca^{2+}、Ba^{2+} 和 Mg^{2+} 实现物理交联。尽管海藻酸盐中单个离子键的能量相对较低，但海藻酸盐链上的多个（超过 20 个）相邻离子交联可以按"蛋盒"模型形成一个密集的交联区域[4,23,85,89]，从而得到相对稳定的海藻酸盐水凝胶。壳聚糖作为阳离子聚电解质，可以与柠檬酸盐、三聚磷酸盐等多价阴离子交联[90-92]。此外，带相反电荷的聚电解质的静电相互作用也可以产生物理交联的水凝胶。例如，只需将阴离子聚 L- 谷氨酸和阳离子聚 L- 赖氨酸混合

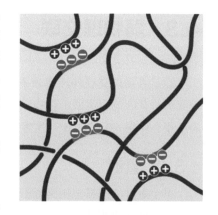

图 4.7 基于静电相互作用实现弱物理交联的示意图

在磷酸盐缓冲盐溶液中即可形成可注射的水凝胶[87]。聚 3-（甲基丙烯酰氨基）丙基 - 三甲基氯化铵和聚对苯乙烯磺酸钠可以形成聚离子络合物，并通过带

相反电荷的单体逐步聚合得到一系列坚韧的自愈性水凝胶[93]。值得注意的是，基于电荷的离子交联的形成，通常需要水凝胶溶剂具有低离子强度，以避免电荷屏蔽。

4.3.3 配位络合

配位络合物由中心金属离子（尤其是过渡金属离子）和周围的有机配体阵列组成（图 4.8）[94,95]。配位键的能量范围为 100～300kJ/mol（图 4.2）[84]。配位键可以在许多组织中（如人骨[96]、昆虫下颌骨[97]、贻贝足丝[98]等）起到结构支撑的作用。配位络合物交联的水凝胶主要由螯合配体与金属离子形成配位络合物，从而对高分子主链进行功能改性而形成。常用的螯合配体有双磷酸盐[99-101]、儿茶酚[102-104]、组氨酸[105-107]、硫醇盐[108,109]、羧酸盐[110,111]、吡啶[112]、联吡啶[113]、亚氨基二乙酸盐[107,114]等，常用的金属离子则有 Cu^{2+}、Zn^{2+}、Fe^{3+}、Co^{2+}、Ni^{2+} 等。双磷酸盐配体可以对透明质酸[115]、明胶[116]、PEG[100] 进行化学修饰，进而与 Ca^{2+}、Mg^{2+}、Ag^+ 等形成配位复合物。除了双磷酸盐外，儿

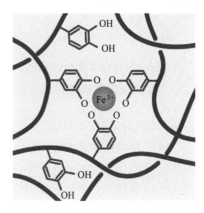

图 4.8 基于配位络合实现弱物理交联的示意图

茶酚配体也被广泛用于 PEG[117,118]、明胶[119]、透明质酸[120]、壳聚糖[121]、聚丙烯酰胺[103]、PAA[104] 等各种功能化高分子。当 pH 值高于 8 时，用 3,4- 二羟基苯基 -1- 丙氨酸（DOPA）残基修饰的 PEG 可以与 Cu^{2+}、Zn^{2+}、Fe^{3+} 等形成配位络合物[78,118]。组氨酸产生的咪唑配体残基[122]，是人体内最重要的螯合剂之一[123]。用组氨酸修饰的 PEG 可以与 Cu^{2+}、Co^{2+}、Ni^{2+} 等形成配位络合物，从而实现 PEG 水凝胶的物理交联[78,124]。此外，通过改变金属离子和 / 或螯合配体还可以调节配位络合物交联的水凝胶的力学性能[105,125]。

4.3.4 主客体相互作用

主客体相互作用是指两个或多个分子或离子通过共价键以外的力以独特的结构关系结合在一起（图 4.9）[126-128]。拥有这种物理作用的两种最常见的主体分子是环糊精和葫芦[n]脲。环糊精（CD）是由 6～8 个 D- 葡萄糖重复单元组成的环状低聚糖，这些重复单元通过 α-(1 → 4)- 糖苷键连接[129,130]。常见的 CD 包括 α-、β- 和 γ-CD，它们分别由 6 个，7 个和 8 个 D- 葡萄糖重复单元组成。这些 CD 呈截锥形，在暴露于溶剂的较小锥缘上拥有仲羟基和伯羟基，这使得 CD 的内腔

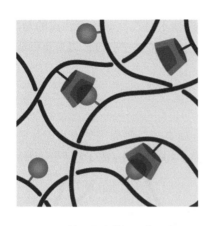

图 4.9 基于主客体相互作用实现弱物理交联的示意图

展现出疏水性，而外表面展现出亲水性[128]。因此，这些环糊精可以通过疏水相互作用和范德华力，充当各种具有适当分子大小的疏水客体分子的主体分子[6,126,131]。例如，α-CD 的常见客体包括偶氮苯[132]和二茂铁[133]；β-CD 的常见客体包括金刚烷[134]、苯并咪唑[135]、3-(三甲基甲硅烷基)丙酸[136]、偶氮苯[132]、二茂铁[137]、联吡啶[138]、酚酞[139]、胆固醇[140]等；γ-CD 的常见客体则是二茂铁[133]。其中，金刚烷由于其与 β-CD 尺寸的互补性以及自身的高疏水性而成为具有最大亲和力的客体分子之一[141]。此外，偶氮苯、二茂铁与 CD 的络合分别对光[142]或氧化还原条件[143,144]有响应。

葫芦[n]脲（CB[n]，n=5～8）是由甲醛和甘脲通过缩合反应制成的南瓜状大环低聚物[128,145]。CB[n]通常具有刚性疏水空腔结构，在开口周围有两个相同的亲水性极性羰基。空腔的尺寸范围为 4.4～8.8Å（对于 CB[n]，n=5～8），其开口直径范围为 2.4～6.9Å[146]。CB 与客体分子的络合常数通常大于其他主体分子结构[147]，主要是由于客体分子与 CB[n]刚性负电性内腔之间会形成强电荷偶极子，存在氢键和疏水/亲水相互作用[148-150]。CB[8]还可以对带正电荷和相对较大的客体（如金刚烷衍生物）表现出高络合能力；且由于其空腔大，可以同时容纳两个客体分子，从而形成高度稳定的三元配合物。例如，CB[8]可以与两个 2,6-双（4,5-二氢-1H-咪唑-2-基）萘分子[151]或一个紫精（百草枯）和一个 2,6-二羟基萘分子形成稳定的络合物[152]。

通过主客体相互作用交联的水凝胶通常由客体分子和/或主体分子修饰的高分子网络构成。例如，单体、主体分子、客体分子可以共聚合成通过主客体相互作用交联的高分子网络[153]。主体/客体分子也可以化学接枝到如 PEG、PDMAA、透明质酸、PAA 等高分子的主链或末端，然后添加相应的双官能团基客体/主体交联剂将高分子网络交联[128]。此外，也可以单独合成客体功能化和主体功能化的高分子，这两种高分子的混合物可以产生通过主客体相互作用交联的水凝胶[154,155]。这些具有主客体相互作用的超分子水凝胶已被广泛用于制造刺激响应性水凝胶材料[156]和其他动态组装水凝胶系统[131,157]。

4.3.5 疏水缔合

基于疏水缔合的物理交联依赖于高分子链疏水域的微相分离和聚集

（图 4.10）[78,81]，其能量范围为 0.1 ～ 20kJ/mol（图 4.2）[84]。疏水域可以通过聚合后改性或直接在聚合物链内共聚疏水单体（随机引入或作为嵌段引入均可）[158]。这些修饰通常需要使用非水溶剂、混合溶剂或胶束系统[159,160]。引入疏水域的一个典型例子是合成具有疏水性丙烯酸正烷基酯端嵌段，和 PEG、聚丙烯酰胺（PAM）、PAA 或 PHEMA 高分子中间嵌段的多嵌段共聚物[161-163]。值得注意的是，尽管疏水缔合的能量通常低于晶畴和玻璃状结节，但由于一个疏水缔合位点可以互连多条高分子链，因此疏水缔合也被用作水凝胶中的功能交联[164,165]。

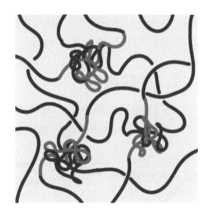

图 4.10 基于疏水缔合实现弱物理交联的示意图

4.3.6 π-π 堆积

π-π 堆积是一种非共价相互作用，特指芳香基团中 π 电子之间的吸引相互作用（图 4.11）[166]。基于芳香族相互作用的几何形状，π-π 堆积可分为边对面（T 形）、偏移和面对面堆叠结构[167]。通常，π-π 堆积的能量范围为 1 ～ 50kJ/mol（图 4.2）[168]。具有芳香环的天然氨基酸（如苯丙氨酸、酪氨酸和色氨酸）以及其他具有共轭结构的化合物（如 1- 芘丁酸、2- 萘乙酸、甲基丙烯酸硝基苯酯

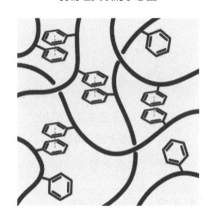

图 4.11 基于 π-π 堆积实现弱物理交联的示意图

等），均可用于设计和制备具有芳香族结构的高分子，并进一步通过 π-π 堆积实现凝胶化[169-171]。例如，含有短肽和 N 端的 Fmoc 氨基酸的芳香族部分可以自组装成稳定的超分子结构[169]。此外，碳纳米管[172,173]、聚噻吩[174,175]、石墨烯基纳米材料[176,177]（包括单层石墨烯、多层石墨烯、氧化石墨烯、还原氧化石墨烯）也能够形成 π-π 堆积结构，适用于导电水凝胶的制备[178,179]。

4.4 动态共价交联

除了弱物理键外，动态共价键也可以作为可逆交联，可受外部刺激裂解。动态共价键的能量通常类似于或低于永久共价键的能量[180]，但高于弱物理键的能

量（图4.2）。水凝胶中动态共价交联的典型例子包括亚胺键、硼酸酯键、二硫键、环己烯腙键、肟键、可逆双烯合成（狄尔斯-阿尔德，Diels-Alder）反应。

4.4.1 亚胺键

亚胺键是碳氮双键，通常由胺与醛或酮之间的反应形成（图4.12）[181]。水凝胶中的亚胺交联通常通过席夫碱反应形成，但会产生脂肪族席夫碱或芳香族席夫碱[182-185]。亚胺交联的可逆性赋予了所得水凝胶机械耗散、自修复和刺激响应等特性[186]。席夫碱的能量范围为67～477kJ/mol（图4.2）[3,187]。一般情况下，芳香族席夫碱比脂肪族席夫碱具有更高的能量和稳定性[188,189]。

由于大多数生物高分子含有氨基，因此酰亚胺键特别适用于制备基于生物高分子的水凝胶。这些胺类可以在温和条件下与各种醛类交联剂反应形成酰亚胺键[186,190]。获得的具有亚胺键的水凝胶通常对pH值、游离胺、游离醛等生化刺激具有响应

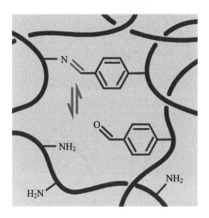

图4.12 基于亚胺键实现动态共价交联的示意图

性[189]。该类水凝胶可用作生物医学应用中的自愈材料和可注射支架[191-193]。

4.4.2 硼酸酯键

动态硼酸酯键，由二醇和硼酸反应形成（图4.13）[194-196]，其能量范围为27.2～93.3kJ/mol（图4.2）[197]，具有pH值和温度依赖性[198-200]。将硼酸基团引入高分子链的方法有两种：一是将硼酸基单体与聚合单体混合共聚，如将含硼酸的单体与其他单体[例如丙烯酰胺（AM）和N-异丙基丙烯酰胺（NIPAM）]一起聚合[201]，即可将硼酸引入水凝胶中[202]；二是化学接枝，如通过碳二亚胺化学将硼酸官能团接枝到预先形成的高分子链上[203,204]。

含硼酸的高分子可以与含有二醇官能团的高分子反应。例如，用硼酸改性的高分子可以在酸性环境中与水杨异羟肟酸基团，或在碱性环境中与邻苯二酚基团，形成动态硼酸酯交

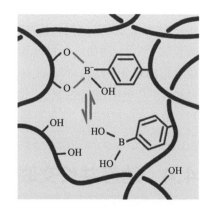

图4.13 基于硼酸酯键实现动态共价交联的示意图

联[117,199,205-207]。此外，PVA[208,209]、海藻酸盐[210,211]、纤维素[212]等多羟基高分子，也可以通过在水溶液中将多羟基聚合物与含硼酸聚合物混合而交联成动态水凝胶。瞬态硼酸酯网络通常可以在断裂后动态重组，使所得水凝胶具有可注射性和自愈性[213,214]。此外，硼酸酯交联水凝胶还对葡萄糖敏感，主要是由于葡萄糖可以与二醇基团竞争形成硼酸酯 - 葡萄糖复合物，从而使水凝胶去交联化[208]。这些基于硼酸酯键的葡萄糖敏感性水凝胶已被用于自我调节胰岛素释放和葡萄糖传感领域[6,203,208,215]。

4.4.3 二硫键

二硫键是在弱碱性环境或温和氧化条件下基于硫醇 - 硫醇相互作用的动态共价键[216,217]，其能量约为425kJ/mol（图4.2）[3,218]。许多天然高分子通过二硫键来稳定结构（图4.14），典型的例子包含纤维蛋白[219]和胶原蛋白[220]。此外，也可以使用含二硫键的交联剂将二硫键引入聚合物中，如3,3′- 二硫代双（丙酸二酰肼）[221,222]和 N,N′- 胱胺 - 双 - 丙烯酰胺[223-225]。由于硫醇 - 硫醇反应具有相对较快的动力学，因此可用于制备动态水凝胶[226,227]。此外，该反应条件相对温和，因而通过二硫键交联的水凝胶可被用于封装各种类型的细胞[228,229]。值得注意的是，二硫键可被三（2- 羧乙基）膦[230]、1,4- 二硫苏糖醇[231]、谷胱甘肽[226,232]等还原剂裂解，因此基于二硫键的水凝胶材料具有氧化还原响应性。

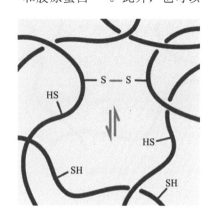

图 4.14 基于二硫键实现动态共价交联的示意图

4.4.4 腙键

腙键由醛基和酰肼基反应形成[233]。PEG[234]、纤维素[235]、多糖[183]等具有羟基的聚合物，可以很容易地用醛和酰肼（或酰肼）基序进行修饰。

由腙键交联的水凝胶可以通过改变pH值表现出可逆的溶胶 - 凝胶转变特性[239-241]。可逆的腙键可通过在生理条件下简单地混合含醛和含酰肼的聚合物形成[236-238]，因此，腙键交联的水凝胶（图4.15）兼具细胞相容性和快速的凝胶化动力学，可用于原位细

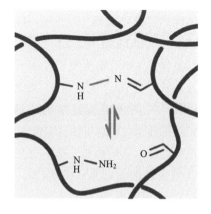

图 4.15 基于腙键实现动态共价交联的示意图

胞封装[242,243]。此外，该类水凝胶的力学性能是可调控的，这有助于研究细胞行为与水凝胶力学（如应力松弛动力学）之间的关系[244]。基于腙键在弱酸性环境（pH 值为 4.0～6.0）中的可逆性，腙键也可用于制备自愈性水凝胶和可注射水凝胶[236,238,240]。

4.4.5 肟键

肟键是由羟胺与醛或酮在温和条件下高效反应形成的[245]。活性醛基或酮基可以通过自由基聚合[246]或氧化反应[247,248]修饰到聚合物上，而羟胺基序主要是通过 N-羟基邻苯二甲酰亚胺诱导的 Mitsunobu 反应和肼还原修饰到含羟基的高分子上[249]。然后，可以通过在中性或弱酸性水溶液中将含醛或酮的聚合物与含羟胺的高分子混合来形成肟键[250]。该反应没有细胞毒性副产物，具有生物相容性，因而可被用于合成基于生物高分子的水凝胶[249,251]。由于动态性质，肟键已被用于构建具有自我修复性能和可注射性的水凝胶，而且该类水凝胶能展现出比亚胺键和腙键交联的水凝胶更高的水解稳定性[250,252]。基于肟键实现动态交联的示意图见图 4.16。

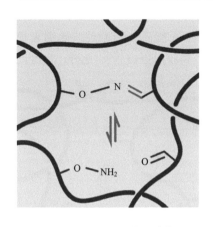

图 4.16　基于肟键实现动态共价交联的示意图

4.4.6 可逆 Diels-Alder 反应

Diels-Alder 反应是二烯和亲二烯基团之间的"点击反应"[253,254]，其能量范围为 37.6～130kJ/mol（图 4.2）[255,256]。可以在天然高分子（如透明质酸[257]、纤维素[258]、其他多糖[259]）和合成高分子（如 PNIPAM[260]、PEG[261]）的主链或链端用二烯（如呋喃）官能团和亲二烯体（如马来酰亚胺）官能团进行修饰，从而构建出基于动态 Diels-Alder 反应交联的水凝胶（图 4.17）。Diels-Alder 连接的平衡具有热响应性。例如，Diels-Alder 加合物在温度升高条件下，会重组成马来酰亚胺和呋喃部分[262,263]。此外，Diels-

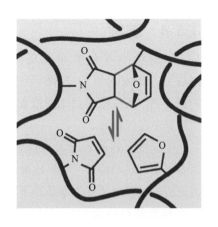

图 4.17　基于可逆 Diels-Alder 反应实现动态共价交联的示意图

Alder 反应可以在生理条件下在水性介质中进行，且不会出现任何副反应或副产物[264-267]。因此，基于 Diels-Alder 反应所构建出的自修复性或自适应性的水凝胶常被用于药物输送和组织工程等领域[257,267,268]。

参考文献

[1] Grandbois M, Beyer M, Rief M, et al. How strong is a covalent bond? Science, 1999, 283 (5408): 1727-1730.

[2] Beyer M K. The mechanical strength of a covalent bond calculated by density functional theory. The Journal of Chemical Physics, 2000, 112 (17): 7307-7312.

[3] Luo Y R. Comprehensive handbook of chemical bond energies. CRC press, 2007.

[4] Lee K Y, Mooney D J. Hydrogels for tissue engineering. Chemical reviews, 2001, 101 (7): 1869-1880.

[5] Ullah F, Othman M B H, Javed F, et al. Classification, processing and application of hydrogels: A review. Materials Science and Engineering: C, 2015, 57: 414-433.

[6] Wang W, Narain R, Zeng H. Rational design of self-healing tough hydrogels: a mini review. Frontiers in chemistry, 2018, 6: 497.

[7] Hennink W E, van Nostrum C F. Novel crosslinking methods to design hydrogels. Advanced drug delivery reviews, 2012, 64: 223-236.

[8] Peppas N A, Merrill E W. Development of semicrystalline poly (vinyl alcohol) hydrogels for biomedical applications. Journal of biomedical materials research, 1977, 11 (3): 423-434.

[9] Wong R S H, Ashton M, Dodou K. Effect of crosslinking agent concentration on the properties of unmedicated hydrogels. Pharmaceutics, 2015, 7 (3): 305-319.

[10] Malo de Molina P, Lad S, Helgeson M E. Heterogeneity and its Influence on the Properties of Difunctional Poly (ethylene glycol) Hydrogels: Structure and Mechanics. Macromolecules, 2015, 48 (15): 5402-5411.

[11] Sawhney A S, Pathak C P, Hubbell J A. Bioerodible hydrogels based on photopolymerized poly (ethylene glycol) -co-poly (.alpha.-hydroxy acid) diacrylate macromers. Macromolecules, 1993, 26 (4): 581-587.

[12] Martens P, Anseth K. Characterization of hydrogels formed from acrylate modified poly (vinyl alcohol) macromers. Polymer, 2000, 41 (21): 7715-7722.

[13] Hubbell J A. Hydrogel systems for barriers and local drug delivery in the control of wound healing. Journal of Controlled Release, 1996, 39 (2-3): 305-313.

[14] van Dijk-Wolthuis W, Franssen O, Talsma H, et al. Synthesis, characterization, and polymerization of glycidyl methacrylate derivatized dextran. Macromolecules, 1995, 28 (18): 6317-6322.

[15] Giammona G, Pitarresi G, Cavallaro G, et al. New biodegradable hydrogels based on an acryloylated polyaspartamide cross-linked by gamma irradiation. Journal of Biomaterials Science, Polymer Edition, 1999, 10 (9): 969-987.

[16] Caliceti P, Salmaso S, Lante A, et al. Controlled release of biomolecules from temperature-sensitive hydrogels prepared by radiation polymerization. Journal of controlled release, 2001, 75 (1-2):

173-181.

[17] Charlesby A. Cross-linking of polythene by pile radiation. Proceedings of the Royal Society of London Series A Mathematical and Physical Sciences, 1952, 215 (1121): 187-214.

[18] Peppas N A, Hydrogels in medicine and pharmacy. CRC press Boca Raton, FL: 1986, 1.

[19] Reddy N, Reddy R, Jiang Q. Crosslinking biopolymers for biomedical applications. Trends in biotechnology, 2015, 33 (6): 362-369.

[20] Kamata H, Akagi Y, Kayasuga-Kariya Y, et al. "Nonswellable" hydrogel without mechanical hysteresis. Science, 2014, 343 (6173): 873-875.

[21] Eiselt P, Lee K Y, Mooney D J. Rigidity of two-component hydrogels prepared from alginate and poly (ethylene glycol) - diamines. Macromolecules, 1999, 32 (17): 5561-5566.

[22] Moon S Y, Jeon E, Bae J S, et al. Polyurea networks via organic sol-gel crosslinking polymerization of tetrafunctional amines and diisocyanates and their selective adsorption and filtration of carbon dioxide. Polymer Chemistry, 2014, 5 (4): 1124-1131.

[23] Hoffman A S. Hydrogels for biomedical applications. Advanced drug delivery reviews, 2012, 64, 18-23.

[24] Meldal M, Tornøe C W. Cu-catalyzed azide- alkyne cycloaddition. Chemical reviews, 2008, 108 (8): 2952-3015.

[25] Agard N J, Prescher J A, Bertozzi C R. A strain-promoted [3+ 2] azide- alkyne cycloaddition for covalent modification of biomolecules in living systems. Journal of the American Chemical Society, 2004, 126 (46): 15046-15047.

[26] Park H, Park K, Shalaby W S, Biodegradable hydrogels for drug delivery. CRC Press, 1993.

[27] Yu Q, Song Y, Shi X, et al. Preparation and properties of chitosan derivative/poly (vinyl alcohol) blend film crosslinked with glutaraldehyde. Carbohydrate Polymers, 2011, 84 (1): 465-470.

[28] Crescenzi V, Francescangeli A, Taglienti A, et al. Synthesis and partial characterization of hydrogels obtained via glutaraldehyde crosslinking of acetylated chitosan and of hyaluronan derivatives. Biomacromolecules, 2003, 4 (4): 1045-1054.

[29] Gehrke S H, Uhden L H, McBride J F. Enhanced loading and activity retention of bioactive proteins in hydrogel delivery systems. Journal of controlled release,1998, 55 (1): 21-33.

[30] Coviello T, Grassi M, Rambone G, et al. Novel hydrogel system from scleroglucan: synthesis and characterization. Journal of controlled release, 1999, 60 (2-3): 367-378.

[31] Tronci G, Neffe A T, Pierce B F, et al. An entropy-elastic gelatin-based hydrogel system. Journal of Materials Chemistry, 2010, 20 (40): 8875-8884.

[32] Rydholm A E, Bowman C N, Anseth K S. Degradable thiol-acrylate photopolymers: polymerization and degradation behavior of an in situ forming biomaterial. Biomaterials, 2005, 26 (22): 4495-4506.

[33] DeForest C A, Polizzotti B D, Anseth K S. Sequential click reactions for synthesizing and patterning three-dimensional cell microenvironments. Nature materials, 2009, 8 (8): 659-664.

[34] Patai S. The chemistry of the thiol group. Wiley, 1974, 2.

[35] Hoyle C E, Lowe A B, Bowman C N. Thiol-click chemistry: a multifaceted toolbox for small molecule and polymer synthesis. Chemical Society Reviews, 2010, 39 (4): 1355-1387.

[36] Gress A, Völkel A, Schlaad H. Thio-click modification of poly [2-(3-butenyl)-2-oxazoline].

Macromolecules, 2007, 40 (22): 7928-7933.

[37] Dondoni A. The emergence of thiol-ene coupling as a click process for materials and bioorganic chemistry. Angewandte Chemie International Edition, 2008, 47 (47): 8995-8997.

[38] Grazú V, Abian O, Mateo C, et al. Novel bifunctional epoxy/thiol-reactive support to immobilize thiol containing proteins by the epoxy chemistry. Biomacromolecules, 2003, 4 (6): 1495-1501.

[39] Dyer E, Glenn J F. The Kinetics of the Reactions of Phenyl Isocyanate with Certain Thiols1. Journal of the American Chemical Society, 1957, 79 (2): 366-369.

[40] Becer C R, Babiuch K, Pilz D, et al. Clicking pentafluorostyrene copolymers: synthesis, nanoprecipitation, and glycosylation. Macromolecules, 2009, 42 (7): 2387-2394.

[41] Hurd C D, Gershbein L L. Reactions of mercaptans with acrylic and methacrylic derivatives. Journal of the American Chemical Society, 1947, 69 (10): 2328-2335.

[42] Chen X, Li R, Wong S H D, et al. Conformational manipulation of scale-up prepared single-chain polymeric nanogels for multiscale regulation of cells. Nature communications, 2019, 10 (1): 2705.

[43] Nair D P, Podgorski M, Chatani S, et al. The thiol-Michael addition click reaction: a powerful and widely used tool in materials chemistry. Chemistry of Materials, 2014, 26 (1): 724-744.

[44] Weinhold F, West R. The nature of the silicon-oxygen bond. Organometallics 2011, 30 (21): 5815-5824.

[45] Tao H, Zhang X, Sun Y, et al. The influence of molecular weight of siloxane macromere on phase separation morphology, oxygen permeability, and mechanical properties in multicomponent silicone hydrogels. Colloid and Polymer Science, 2017, 295: 205-213.

[46] Zhang X, Wang L, Tao H, et al. The influences of poly (ethylene glycol) chain length on hydrophilicity, oxygen permeability, and mechanical properties of multicomponent silicone hydrogels. Colloid and Polymer Science, 2019, 297: 1233-1243.

[47] Si L, Zheng X, Nie J, et al. Silicone-based tough hydrogels with high resilience, fast self-recovery, and self-healing properties. Chemical Communications, 2016, 52 (54): 8365-8368.

[48] Yuk H, Zhang T, Lin S, et al. Tough bonding of hydrogels to diverse non-porous surfaces. Nature materials, 2016, 15 (2): 190-196.

[49] Hassan C M, Peppas N A. Structure and morphology of freeze/thawed PVA hydrogels. Macromolecules, 2000, 33 (7): 2472-2479.

[50] Kumar A, Han S S, PVA-based hydrogels for tissue engineering: A review. International journal of polymeric materials and polymeric biomaterials, 2017, 66 (4): 159-182.

[51] Peppas N A, Hilt J Z, Khademhosseini A, et al. Hydrogels in biology and medicine: from molecular principles to bionanotechnology. Advanced materials, 2006, 18 (11): 1345-1360.

[52] Croisier F, Jérôme C. Chitosan-based biomaterials for tissue engineering. European polymer journal, 2013, 49 (4): 780-792.

[53] Berger J, Reist M, Mayer J M, et al. Structure and interactions in covalently and ionically crosslinked chitosan hydrogels for biomedical applications. European journal of pharmaceutics and biopharmaceutics, 2004, 57 (1): 19-34.

[54] Delmer D P, Amor Y. Cellulose biosynthesis. The plant cell, 1995, 7 (7): 987.

[55] Abe K, Yano H. Formation of hydrogels from cellulose nanofibers. Carbohydrate Polymers, 2011, 85

(4): 733-737.

[56] Abe K, Yano H. Cellulose nanofiber-based hydrogels with high mechanical strength. Cellulose, 2012, 19: 1907-1912.

[57] Liu J, Lin S, Liu X, et al. Fatigue-resistant adhesion of hydrogels. Nature communications, 2020, 11 (1): 1071.

[58] Stillinger F H, Debenedetti P. Supercooled liquids and the glass transition. Nature, 2001, 410: 259-267.

[59] Burns A B, Register R A. Mechanical properties of star block polymer thermoplastic elastomers with glassy and crystalline end blocks. Macromolecules, 2016, 49 (24): 9521-9530.

[60] Nykänen A, Nuopponen M, Laukkanen A, et al. Phase Behavior and Temperature-Responsive Molecular Filters Based on Self-Assembly of Polystyrene-block-poly (N-isopropylacrylamide) -block-polystyrene. Macromolecules, 2007, 40 (16): 5827-5834.

[61] Roth C, Pound A, Kamp S, et al. Molecular-weight dependence of the glass transition temperature of freely-standing poly (methyl methacrylate) films. The European Physical Journal E, 2006, 20: 441-448.

[62] Seitz M E, Martina D, Baumberger T, et al. Fracture and large strain behavior of self-assembled triblock copolymer gels. Soft Matter, 2009, 5 (2): 447-456.

[63] Yan C, Pochan D J. Rheological properties of peptide-based hydrogels for biomedical and other applications. Chemical Society Reviews, 2010, 39 (9): 3528-3540.

[64] Viebke C, Piculell L, Nilsson S. On the mechanism of gelation of helix-forming biopolymers. Macromolecules, 1994, 27 (15): 4160-4166.

[65] Prince E, Kumacheva E. Design and applications of man-made biomimetic fibrillar hydrogels. Nature Reviews Materials, 2019, 4 (2): 99-115.

[66] Ottani V, Martini D, Franchi M, et al. Hierarchical structures in fibrillar collagens. Micron, 2002, 33 (7-8): 587-596.

[67] O'leary L E, Fallas J A, Bakota E L, et al. Multi-hierarchical self-assembly of a collagen mimetic peptide from triple helix to nanofibre and hydrogel. Nature chemistry, 2011, 3 (10): 821-828.

[68] Arnott S, Fulmer A, Scott W, et al. The agarose double helix and its function in agarose gel structure. Journal of molecular biology, 1974, 90 (2): 269-284.

[69] Foord S, Atkins E. New X-ray diffraction results from agarose: Extended single helix structures and implications for gelation mechanism. Biopolymers: Original Research on Biomolecules, 1989, 28 (8): 1345-1365.

[70] Armisén R. In Agar and agarose biotechnological applications, International Workshop on Gelidium: Proceedings of the International Workshop on Gelidium held in Santander, Spain, September 3-8, 1990, Springer, 1991: 157-166.

[71] Steiner T. The hydrogen bond in the solid state. Angewandte Chemie International Edition, 2002, 41 (1): 48-76.

[72] Markovitch O, Agmon N. Structure and energetics of the hydronium hydration shells. The Journal of Physical Chemistry A, 2007, 111 (12): 2253-2256.

[73] Duconseille A, Astruc T, Quintana N, et al. Gelatin structure and composition linked to hard capsule

dissolution: A review. Food hydrocolloids, 2015, 43: 360-376.

[74] Van Vlierberghe S, Dubruel P, Schacht E. Biopolymer-based hydrogels as scaffolds for tissue engineering applications: a review. Biomacromolecules, 2011, 12 (5): 1387-1408.

[75] Stauffer S R, Peppast N A. Poly (vinyl alcohol) hydrogels prepared by freezing-thawing cyclic processing. Polymer, 1992, 33 (18): 3932-3936.

[76] Eagland D, Crowther N, Butler C. Complexation between polyoxyethylene and polymethacrylic acid—the importance of the molar mass of polyoxyethylene. European polymer journal, 1994, 30 (7): 767-773.

[77] Kim I S, Kim S H, Cho C S. Drug release from pH-sensitive interpenetrating polymer networks hydrogel based on poly (ethylene glycol) macromer and poly (acrylic acid) prepared by UV cured method. Archives of Pharmacal Research, 1996, 19: 18-22.

[78] Seiffert S, Kumacheva E, Okay O, et al. Supramolecular polymer networks and gels. Springer, 2015, 268.

[79] Voorhaar L, Hoogenboom R. Supramolecular polymer networks: hydrogels and bulk materials. Chemical Society Reviews, 2016, 45 (14): 4013-4031.

[80] Zhang G, Chen Y, Deng Y, et al. Dynamic supramolecular hydrogels: regulating hydrogel properties through self-complementary quadruple hydrogen bonds and thermo-switch. ACS Macro Letters, 2017, 6 (7): 641-646.

[81] Jiang H, Duan L, Ren X, et al. Hydrophobic association hydrogels with excellent mechanical and self-healing properties. European Polymer Journal, 2019, 112: 660-669.

[82] Cui J, del Campo A. Multivalent H-bonds for self-healing hydrogels. Chemical Communications, 2012, 48 (74): 9302-9304.

[83] Li F, Tang J, Geng J, et al. Polymeric DNA hydrogel: Design, synthesis and applications. Progress in Polymer Science, 2019, 98: 101163.

[84] Evans B. The Wiley-Blackwell Encyclopedia of Globalization, 2012.

[85] Lee K Y, Mooney D J. Alginate: properties and biomedical applications. Progress in polymer science, 2012, 37 (1): 106-126.

[86] Sun J Y, Zhao X, Illeperuma W R, et al. Highly stretchable and tough hydrogels. Nature, 2012, 489 (7414): 133-136.

[87] Cui H, Zhuang X, He C, et al. High performance and reversible ionic polypeptide hydrogel based on charge-driven assembly for biomedical applications. Acta biomaterialia, 2015, 11: 183-190.

[88] Ji D Y, Kuo T F, Wu H D, et al. A novel injectable chitosan/polyglutamate polyelectrolyte complex hydrogel with hydroxyapatite for soft-tissue augmentation. Carbohydrate polymers, 2012, 89 (4): 1123-1130.

[89] Lee K Y, Rowley J A, Eiselt P, et al. Controlling mechanical and swelling properties of alginate hydrogels independently by cross-linker type and cross-linking density. Macromolecules, 2000, 33 (11): 4291-4294.

[90] Yang Y, Wang X, Yang F, et al. Highly elastic and ultratough hybrid ionic-covalent hydrogels with tunable structures and mechanics. Advanced Materials, 2018, 30 (18): 1707071.

[91] Pakdel P M, Peighambardoust S J. Review on recent progress in chitosan-based hydrogels for

wastewater treatment application. Carbohydrate polymers, 2018, 201: 264-279.

[92] Mi F L, Sung H W, Shyu S S, et al. Synthesis and characterization of biodegradable TPP/genipin co-crosslinked chitosan gel beads. Polymer, 2003, 44 (21): 6521-6530.

[93] Luo F, Sun T L, Nakajima T, et al. Oppositely charged polyelectrolytes form tough, self-healing, and rebuildable hydrogels. Advanced materials, 2015, 27 (17): 2722-2727.

[94] Mozhdehi D, Ayala S, Cromwell O R, et al. Self-healing multiphase polymers via dynamic metal-ligand interactions. Journal of the American Chemical Society, 2014, 136 (46): 16128-16131.

[95] Zhang J, Su C Y. Metal-organic gels: From discrete metallogelators to coordination polymers. Coordination Chemistry Reviews, 2013, 257 (7-8): 1373-1408.

[96] Fantner G E, Hassenkam T, Kindt J H, et al. Sacrificial bonds and hidden length dissipate energy as mineralized fibrils separate during bone fracture. Nature materials, 2005, 4 (8): 612-616.

[97] Hillerton J E, Vincent J F. The specific location of zinc in insect mandibles, 1982.

[98] Harrington M J, Masic A, Holten-Andersen N, et al. Iron-clad fibers: a metal-based biological strategy for hard flexible coatings. Science, 2010, 328 (5975): 216-220.

[99] Nejadnik M R, Yang X, Bongio M, et al. Self-healing hybrid nanocomposites consisting of bisphosphonated hyaluronan and calcium phosphate nanoparticles. Biomaterials, 2014, 35 (25): 6918-6929.

[100] Lopez-Perez P M, da Silva R M, Strehin I, et al. Self-healing hydrogels formed by complexation between calcium ions and bisphosphonate-functionalized star-shaped polymers. Macromolecules, 2017, 50 (21): 8698-8706.

[101] Shi L, Carstensen H, Hölzl K, et al. Dynamic coordination chemistry enables free directional printing of biopolymer hydrogel. Chemistry of Materials, 2017, 29 (14): 5816-5823.

[102] Li Q, Barrett D G, Messersmith P B, et al. Controlling hydrogel mechanics via bio-inspired polymer-nanoparticle bond dynamics. ACS nano, 2016, 10 (1): 1317-1324.

[103] Hou S, Ma P X. Stimuli-responsive supramolecular hydrogels with high extensibility and fast self-healing via precoordinated mussel-inspired chemistry. Chemistry of Materials, 2015, 27 (22): 7627-7635.

[104] Wang W, Xu Y, Li A, et al. Zinc induced polyelectrolyte coacervate bioadhesive and its transition to a self-healing hydrogel. Rsc Advances, 2015, 5 (82): 66871-66878.

[105] Grindy S C, Learsch R, Mozhdehi D, et al. Control of hierarchical polymer mechanics with bioinspired metal-coordination dynamics. Nature materials, 2015, 14 (12): 1210-1216.

[106] Fullenkamp D E, He L, Barrett D G, et al. Mussel-inspired histidine-based transient network metal coordination hydrogels. Macromolecules, 2013, 46 (3): 1167-1174.

[107] Wang C, Stewart R J, Kopeček J. Hybrid hydrogels assembled from synthetic polymers and coiled-coil protein domains. Nature, 1999, 397 (6718): 417-420.

[108] Qin H, Zhang T, Li H N, et al. Dynamic Au-thiolate interaction induced rapid self-healing nanocomposite hydrogels with remarkable mechanical behaviors. Chem, 2017, 3 (4): 691-705.

[109] Casuso P, Odriozola I, Pérez-San Vicente A, et al. Injectable and self-healing dynamic hydrogels based on metal (Ⅰ) -thiolate/disulfide exchange as biomaterials with tunable mechanical properties. Biomacromolecules, 2015, 16 (11): 3552-3561.

[110] Peng F, Li G, Liu X, et al. Redox-responsive gel- sol/sol- gel transition in poly (acrylic acid) aqueous solution containing Fe (Ⅲ) ions switched by light. Journal of the American Chemical Society, 2008, 130 (48): 16166-16167.

[111] Zheng S Y, Ding H, Qian J, et al. Metal-coordination complexes mediated physical hydrogels with high toughness, stick-slip tearing behavior, and good processability. Macromolecules, 2016, 49 (24): 9637-9646.

[112] Buwalda S J, Dijkstra P J, Feijen J. Poly (ethylene glycol) -poly (l-lactide) star block copolymer hydrogels crosslinked by metal-ligand coordination. Journal of Polymer Science Part A: Polymer Chemistry, 2012, 50 (9): 1783-1791.

[113] Chujo Y, Sada K, Saegusa T. Iron (Ⅱ) bipyridyl-branched polyoxazoline complex as a thermally reversible hydrogel. Macromolecules, 1993, 26 (24): 6315-6319.

[114] Weng G, Thanneeru S, He J. Dynamic coordination of Eu-iminodiacetate to control fluorochromic response of polymer hydrogels to multistimuli. Advanced Materials, 2018, 30 (11): 1706526.

[115] Zhang K, Feng Q, Xu J, et al. Self-Assembled injectable nanocomposite hydrogels stabilized by bisphosphonate-magnesium (Mg^{2+}) coordination regulates the differentiation of encapsulated stem cells via dual crosslinking. Advanced Functional Materials, 2017, 27 (34): 1701642.

[116] Diba M, Camargo W A, Brindisi M, et al. Composite colloidal gels made of bisphosphonate-functionalized gelatin and bioactive glass particles for regeneration of osteoporotic bone defects. Advanced Functional Materials, 2017, 27 (45): 1703438.

[117] He L, Fullenkamp D E, Rivera J G, et al. pH responsive self-healing hydrogels formed by boronate-catechol complexation. Chemical Communications, 2011, 47 (26): 7497-7499.

[118] Holten-Andersen N, Harrington M J, Birkedal H, et al. pH-induced metal-ligand cross-links inspired by mussel yield self-healing polymer networks with near-covalent elastic moduli. Proceedings of the National Academy of Sciences, 2011, 108 (7): 2651-2655.

[119] Choi Y C, Choi J S, Jung Y J, et al. Human gelatin tissue-adhesive hydrogels prepared by enzyme-mediated biosynthesis of DOPA and Fe 3+ ion crosslinking. Journal of Materials Chemistry B, 2014, 2 (2): 201-209.

[120] Lee J, Chang K, Kim S, et al. Phase controllable hyaluronic acid hydrogel with iron (Ⅲ) ion-catechol induced dual cross-linking by utilizing the gap of gelation kinetics. Macromolecules, 2016, 49 (19): 7450-7459.

[121] Yavvari P S, Srivastava A. Robust, self-healing hydrogels synthesised from catechol rich polymers. Journal of Materials Chemistry B, 2015, 3 (5): 899-910.

[122] Gent A N. A new constitutive relation for rubber. Rubber chemistry and technology, 1996, 69 (1): 59-61.

[123] Sundberg R J, Martin R B. Interactions of histidine and other imidazole derivatives with transition metal ions in chemical and biological systems. Chemical reviews, 1974, 74 (4): 471-517.

[124] Shi L, Ding P, Wang Y, et al. Self-healing polymeric hydrogel formed by metal-ligand coordination assembly: design, fabrication, and biomedical applications. Macromolecular rapid communications, 2019, 40 (7): 1800-1837.

[125] Cook T R, Zheng Y R, Stang P J. Metal-organic frameworks and self-assembled supramolecular

coordination complexes: comparing and contrasting the design, synthesis, and functionality of metal-organic materials. Chemical reviews, 2013, 113 (1): 734-777.

[126] Harada A, Takashima Y, Yamaguchi H. Cyclodextrin-based supramolecular polymers. Chemical Society Reviews, 2009, 38 (4): 875-882.

[127] Rodell C B, Mealy J E, Burdick J A. Supramolecular guest-host interactions for the preparation of biomedical materials. Bioconjugate chemistry, 2015, 26 (12): 2279-2289.

[128] Appel E A, del Barrio J, Loh X J, et al. Supramolecular polymeric hydrogels. Chemical Society Reviews, 2012, 41 (18): 6195-6214.

[129] Szejtli J. Introduction and general overview of cyclodextrin chemistry. Chemical reviews, 1998, 98 (5): 1743-1754.

[130] Harada A, Takashima Y, Nakahata M. Supramolecular polymeric materials via cyclodextrin-guest interactions. Accounts of chemical research, 2014, 47 (7): 2128-2140.

[131] Liu G, Yuan Q, Hollett G, et al. Cyclodextrin-based host-guest supramolecular hydrogel and its application in biomedical fields. Polymer Chemistry, 2018, 9 (25): 3436-3449.

[132] Yamaguchi H, Kobayashi Y, Kobayashi R, et al. Photoswitchable gel assembly based on molecular recognition. Nature communications, 2012, 3 (1): 603.

[133] Wu J S, Toda K, Tanaka A, et al. Association constants of ferrocene with cyclodextrins in aqueous medium determined by solubility measurements of ferrocene. Bulletin of the Chemical Society of Japan, 1998, 71 (7): 1615-1618.

[134] Voskuhl J, Waller M, Bandaru S, et al. Nanodiamonds in sugar rings: an experimental and theoretical investigation of cyclodextrin-nanodiamond inclusion complexes. Organic & Biomolecular Chemistry, 2012, 10 (23): 4524-4530.

[135] Lezcano M, Al-Soufi W, Novo M, et al. Complexation of several benzimidazole-type fungicides with α-and β-cyclodextrins. Journal of agricultural and food chemistry, 2002, 50 (1): 108-112.

[136] Huh K M, Tomita H, Lee W K, et al. Synthesis of α-Cyclodextrin-Conjugated Poly (ε-lysine) s and Their Inclusion Complexation Behavior. Macromolecular rapid communications, 2002, 23 (3): 179-182.

[137] Tomatsu I, Hashidzume A, Harada A. Redox-responsive hydrogel system using the molecular recognition of β-cyclodextrin. Macromolecular rapid communications, 2006, 27 (4): 238-241.

[138] Nelissen H F, Feiters M C, Nolte R J. Synthesis and self-inclusion of bipyridine-spaced cyclodextrin dimers. The Journal of Organic Chemistry, 2002, 67 (17): 5901-5906.

[139] Hetzer M, Fleischmann C, Schmidt B V, et al. Visual recognition of supramolecular graft polymer formation via phenolphthalein-cyclodextrin association. Polymer, 2013, 54 (19): 5141-5147.

[140] Ravichandran R, Divakar S. Inclusion of ring A of cholesterol inside the β-cyclodextrin cavity: evidence from oxidation reactions and structural studies. Journal of inclusion phenomena and molecular recognition in chemistry, 1998, 30: 253-270.

[141] Rekharsky M V, Inoue Y. Complexation thermodynamics of cyclodextrins. Chemical reviews, 1998, 98 (5): 1875-1918.

[142] Bortolus P, Monti S. Cis dblharw trans Photoisomerization of azobenzene-cyclodextrin inclusion complexes. Journal of Physical Chemistry, 1987, 91 (19): 5046-5050.

[143] Matsue T, Evans D H, Osa T, et al. Electron-transfer reactions associated with host-guest complexation. Oxidation of ferrocenecarboxylic acid in the presence of. beta.-cyclodextrin. Journal of the American Chemical Society, 1985, 107 (12): 3411-3417.

[144] Kaifer A E. Interplay between molecular recognition and redox chemistry. Accounts of chemical research, 1999, 32 (1): 62-71.

[145] Shih N y. Host-guest chemistry of cucurbituril. University of Illinois at Chicago, 1981.

[146] Isaacs L, Lagona J, Mukhopadhyay P, et al. The cucurbit [n] uril family. Angew. Chem. Int. Ed, 2005, 44: 4844-4870.

[147] Liu S, Ruspic C, Mukhopadhyay P, et al. The cucurbit [n] uril family: prime components for self-sorting systems. Journal of the American Chemical Society, 2005, 127 (45): 15959-15967.

[148] Mock W L, Shih N Y, Structure and selectivity in host-guest complexes of cucurbituril. The Journal of Organic Chemistry, 1986, 51 (23): 4440-4446.

[149] Jeon W S, Moon K, Park S H, et al. Complexation of ferrocene derivatives by the cucurbit [7] uril host: a comparative study of the cucurbituril and cyclodextrin host families. Journal of the American Chemical Society, 2005, 127 (37): 12984-12989.

[150] Moghaddam S, Yang C, Rekharsky M, et al. New ultrahigh affinity host- guest complexes of cucurbit [7] uril with bicyclo [2.2.2] octane and adamantane guests: Thermodynamic analysis and evaluation of m2 affinity calculations. Journal of the American Chemical Society, 2011, 133 (10): 3570-3581.

[151] Kim J, Jung I S, Kim S Y, et al. New cucurbituril homologues: syntheses, isolation, characterization, and X-ray crystal structures of cucurbit [n] uril (n=5, 7, and 8). Journal of the American Chemical Society, 2000, 122 (3): 540-541.

[152] Lee J W, Kim K, Choi S, et al. Unprecedented host-induced intramolecular charge-transfer complex formation. Chemical communications, 2002 (22): 2692-2693.

[153] Takashima Y, Sawa Y, Iwaso K, et al. Supramolecular materials cross-linked by host-guest inclusion complexes: the effect of side chain molecules on mechanical properties. Macromolecules, 2017, 50 (8): 3254-3261.

[154] Jin J, Cai L, Jia Y G, et al. Progress in self-healing hydrogels assembled by host-guest interactions: preparation and biomedical applications. Journal of Materials Chemistry B, 2019, 7 (10): 1637-1651.

[155] Yang X, Yu H, Wang L, et al. Self-healing polymer materials constructed by macrocycle-based host-guest interactions. Soft Matter, 2015, 11 (7): 1242-1252.

[156] Takashima Y, Hatanaka S, Otsubo M, et al. Expansion-contraction of photoresponsive artificial muscle regulated by host-guest interactions. Nature communications, 2012, 3 (1): 1270.

[157] Zhou L, Li J, Luo Q, et al. Dual stimuli-responsive supramolecular pseudo-polyrotaxane hydrogels. Soft Matter, 2013, 9 (18): 4635-4641.

[158] Jeong B, Kim S W, Bae Y H. Thermosensitive sol-gel reversible hydrogels. Advanced drug delivery reviews, 2012, 64: 154-162.

[159] Mihajlovic M, Staropoli M, Appavou M S, et al. Tough supramolecular hydrogel based on strong hydrophobic interactions in a multiblock segmented copolymer. Macromolecules, 2017, 50 (8):

3333-3346.

[160] Tuncaboylu D C, Sari M, Oppermann W, et al. Tough and self-healing hydrogels formed via hydrophobic interactions. Macromolecules, 2011, 44 (12): 4997-5005.

[161] Patrickios C S, Georgiou T K. Covalent amphiphilic polymer networks. Current opinion in colloid & interface science, 2003, 8 (1): 76-85.

[162] Klaikherd A, Nagamani C, Thayumanavan S. Multi-stimuli sensitive amphiphilic block copolymer assemblies. Journal of the American Chemical Society, 2009, 131 (13): 4830-4838.

[163] Zhang H J, Sun T L, Zhang A K, et al. Tough physical double-network hydrogels based on amphiphilic triblock copolymers. Advanced Materials, 2016, 28 (24): 4884-4890.

[164] Gulyuz U, Okay O. Self-healing poly (acrylic acid) hydrogels with shape memory behavior of high mechanical strength. Macromolecules, 2014, 47 (19): 6889-6899.

[165] Tuncaboylu D C, Argun A, Sahin M, et al. Structure optimization of self-healing hydrogels formed via hydrophobic interactions. Polymer, 2012, 53 (24): 5513-5522.

[166] Wheeler S E. Understanding substituent effects in noncovalent interactions involving aromatic rings. Accounts of chemical research, 2013, 46 (4): 1029-1038.

[167] Butterfield S M, Patel P R, Waters M L. Contribution of aromatic interactions to α-helix stability. Journal of the American Chemical Society, 2002, 124 (33): 9751-9755.

[168] Zhuang W R, Wang Y, Cui P F, et al. Applications of π-π stacking interactions in the design of drug-delivery systems. Journal of controlled release, 2019, 294: 311-326.

[169] Ma M, Kuang Y, Gao Y, et al. Aromatic-aromatic interactions induce the self-assembly of pentapeptidic derivatives in water to form nanofibers and supramolecular hydrogels. Journal of the American Chemical Society, 2010, 132 (8): 2719-2728.

[170] Singh V, Snigdha K, Singh C, et al. Understanding the self-assembly of Fmoc-phenylalanine to hydrogel formation. Soft Matter, 2015, 11 (26): 5353-5364.

[171] Yan X, Zhu P, Li J. Self-assembly and application of diphenylalanine-based nanostructures. Chemical Society Reviews, 2010, 39 (6): 1877-1890.

[172] Mehra N K, Palakurthi S. Interactions between carbon nanotubes and bioactives: a drug delivery perspective. Drug Discovery Today, 2016, 21 (4): 585-597.

[173] Kovtyukhova N I, Mallouk T E, Pan L, et al. Individual single-walled nanotubes and hydrogels made by oxidative exfoliation of carbon nanotube ropes. Journal of the American Chemical Society, 2003, 125 (32): 9761-9769.

[174] Lu B, Yuk H, Lin S, et al. Pure pedot: Pss hydrogels. Nature communications, 2019, 10 (1): 1043.

[175] Yuk H, Lu B, Lin S, et al. 3D printing of conducting polymers. Nature communications, 2020, 11 (1): 1604.

[176] Ghawanmeh A A, Ali G A, Algarni H, et al. Graphene oxide-based hydrogels as a nanocarrier for anticancer drug delivery. Nano Research, 2019, 12: 973-990.

[177] Loh K P, Bao Q, Ang P K, et al. The chemistry of graphene. Journal of Materials Chemistry, 2010, 20 (12): 2277-2289.

[178] Yuk H, Lu B, Zhao X. Hydrogel bioelectronics. Chemical Society Reviews, 2019, 48 (6): 1642-1667.

[179] Sinha A, Kalambate P K, Mugo S M, et al. Polymer hydrogel interfaces in electrochemical sensing strategies: A review. TrAC Trends in Analytical Chemistry, 2019, 118: 488-501.

[180] Krishnakumar B, Sanka R P, Binder W H, et al. Vitrimers: Associative dynamic covalent adaptive networks in thermoset polymers. Chemical Engineering Journal, 2020, 385: 123820.

[181] Cordes E H, Jencks W P. The mechanism of hydrolysis of Schiff bases derived from aliphatic amines. Journal of the American Chemical Society, 1963, 85 (18): 2843-2848.

[182] Lü S, Gao C, Xu X, et al. Injectable and self-healing carbohydrate-based hydrogel for cell encapsulation. ACS applied materials & interfaces, 2015, 7 (23): 13029-13037.

[183] Wei Z, Yang J H, Liu Z Q, et al. Novel biocompatible polysaccharide-based self-healing hydrogel. Advanced Functional Materials, 2015, 25 (9): 1352-1359.

[184] Karimi A R, Khodadadi A. Mechanically robust 3D nanostructure chitosan-based hydrogels with autonomic self-healing properties. ACS Applied Materials & Interfaces, 2016, 8 (40): 27254-27263.

[185] Wu X, He C, Wu Y, et al. Synergistic therapeutic effects of Schiff's base cross-linked injectable hydrogels for local co-delivery of metformin and 5-fluorouracil in a mouse colon carcinoma model. Biomaterials, 2016, 75: 148-162.

[186] Ding F, Wu S, Wang S, et al. A dynamic and self-crosslinked polysaccharide hydrogel with autonomous self-healing ability. Soft Matter, 2015, 11 (20): 3971-3976.

[187] Blanco F, Alkorta I, Elguero J. Barriers about double carbon-nitrogen bond in imine derivatives (aldimines, oximes, hydrazones, azines). Croatica Chemica Acta, 2009, 82 (1): 173-183.

[188] Engel A K, Yoden T, Sanui K, et al. Synthesis of aromatic Schiff base oligomers at the air/water interface. Journal of the American Chemical Society, 1985, 107 (26): 8308-8310.

[189] Zhang Y, Tao L, Li S, et al. Synthesis of multiresponsive and dynamic chitosan-based hydrogels for controlled release of bioactive molecules. Biomacromolecules, 2011, 12 (8): 2894-2901.

[190] Huang W, Wang Y, Huang Z, et al. On-demand dissolvable self-healing hydrogel based on carboxymethyl chitosan and cellulose nanocrystal for deep partial thickness burn wound healing. ACS applied materials & interfaces, 2018, 10 (48): 41076-41088.

[191] Boehnke N, Cam C, Bat E, et al. Imine hydrogels with tunable degradability for tissue engineering. Biomacromolecules, 2015, 16 (7): 2101-2108.

[192] Xu Y, Li Y, Chen Q, et al. Injectable and self-healing chitosan hydrogel based on imine bonds: design and therapeutic applications. International Journal of Molecular Sciences, 2018, 19 (8): 2198.

[193] Wang H, Heilshorn S C. Adaptable hydrogel networks with reversible linkages for tissue engineering. Advanced Materials, 2015, 27 (25): 3717-3736.

[194] Bull S D, Davidson M G, Van den Elsen J M, et al. Exploiting the reversible covalent bonding of boronic acids: recognition, sensing, and assembly. Accounts of chemical research, 2013, 46 (2): 312-326.

[195] Guan Y, Zhang Y. Boronic acid-containing hydrogels: synthesis and their applications. Chemical Society Reviews, 2013, 42 (20): 8106-8121.

[196] Nishiyabu R, Kubo Y, James T D, et al. Boronic acid building blocks: tools for self assembly.

Chemical Communications, 2011, 47 (4): 1124-1150.

[197] Li H, Li H, Dai Q, et al. Hydrolytic Stability of Boronate Ester-Linked Covalent Organic Frameworks. Advanced Theory and Simulations, 2018, 1 (2): 1700015.

[198] Pizer R, Babcock L. Mechanism of the complexation of boron acids with catechol and substituted catechols. Inorganic Chemistry, 1977, 16 (7): 1677-1681.

[199] Jay J I, Shukair S, Langheinrich K, et al. Modulation of viscoelasticity and HIV transport as a function of pH in a reversibly crosslinked hydrogel. Advanced functional materials, 2009, 19 (18): 2969-2977.

[200] Mahalingam A, Jay J I, Langheinrich K, et al. Inhibition of the transport of HIV in vitro using a pH-responsive synthetic mucin-like polymer system. Biomaterials, 2011, 32 (33): 8343-8355.

[201] Lowe C, Sartain F, Yang X. Holographic Lactate Sensor. Analytical Chemistry, 2006, 78: 5664-5670.

[202] Kazunori K, Hiroaki M, Masayuki B, et al. Totally synthetic polymer gels responding to external glucose concentration: their preparation and application to on- off regulation of insulin release. Journal of the American Chemical Society, 1998, 120 (48): 12694-12695.

[203] Asher S A, Alexeev V L, Goponenko A V, et al. Photonic crystal carbohydrate sensors: low ionic strength sugar sensing. Journal of the American Chemical Society, 2003, 125 (11): 3322-3329.

[204] Zhang Y, Guan Y, Zhou S. Synthesis and volume phase transitions of glucose-sensitive microgels. Biomacromolecules, 2006, 7 (11): 3196-3201.

[205] Roberts M C, Hanson M C, Massey A P, et al. Dynamically restructuring hydrogel networks formed with reversible covalent crosslinks. Advanced Materials, 2007, 19 (18): 2503-2507.

[206] Nakahata M, Mori S, Takashima Y, et al. pH-and sugar-responsive gel assemblies based on boronate-catechol interactions. ACS Macro Letters, 2014, 3 (4): 337-340.

[207] Shan M, Gong C, Li B, et al. A pH, glucose, and dopamine triple-responsive, self-healable adhesive hydrogel formed by phenylborate-catechol complexation. Polymer Chemistry, 2017, 8 (19): 2997-3005.

[208] Kitano S, Hisamitsu I, Koyama Y, et al. Effect of the incorporation of amino groups in a glucose-responsive polymer complex having phenylboronic acid moieties. Polymers for Advanced Technologies, 1991, 2 (5): 261-264.

[209] Ivanov A, Larsson H, Galaev I Y, et al. Synthesis of boronate-containing copolymers of N, N-dimethylacrylamide, their interaction with poly (vinyl alcohol) and rheological behaviour of the gels. Polymer, 2004, 45 (8): 2495-2505.

[210] Hong S H, Kim S, Park J P, et al. Dynamic bonds between boronic acid and alginate: hydrogels with stretchable, self-healing, stimuli-responsive, remoldable, and adhesive properties. Biomacromolecules, 2018, 19 (6): 2053-2061.

[211] Pettignano A, Grijalvo S, Haering M, et al. Boronic acid-modified alginate enables direct formation of injectable, self-healing and multistimuli-responsive hydrogels. Chemical communications, 2017, 53 (23): 3350-3353.

[212] An H, Bo Y, Chen D, et al. Cellulose-based self-healing hydrogel through boronic ester bonds with excellent biocompatibility and conductivity. RSC advances, 2020, 10 (19): 11300-11310.

[213] Jay J I, Langheinrich K, Hanson M C, et al. Unequal stoichiometry between crosslinking moieties affects the properties of transient networks formed by dynamic covalent crosslinks. Soft Matter, 2011, 7 (12): 5826-5835.

[214] Phadke A, Zhang C, Arman B, et al. Rapid self-healing hydrogels. Proceedings of the National Academy of Sciences, 2012, 109 (12): 4383-4388.

[215] Deng C C, Brooks W L, Abboud K A, et al. Boronic acid-based hydrogels undergo self-healing at neutral and acidic pH. ACS Macro Letters, 2015, 4 (2): 220-224.

[216] Patenaude M, Smeets N M, Hoare T. Designing injectable, covalently cross-linked hydrogels for biomedical applications. Macromolecular rapid communications, 2014, 35 (6): 598-617.

[217] Wilson J M, Bayer R J, Hupe D. Structure-reactivity correlations for the thiol-disulfide interchange reaction. Journal of the American Chemical Society, 1977, 99 (24): 7922-7926.

[218] Dopieralski P, Ribas-Arino J, Anjukandi P, et al. Unexpected mechanochemical complexity in the mechanistic scenarios of disulfide bond reduction in alkaline solution. Nature Chemistry, 2017, 9 (2): 164-170.

[219] Madrazo J, Brown J H, Litvinovich S, et al. Crystal structure of the central region of bovine fibrinogen (E5 fragment) at 1.4-Å resolution. Proceedings of the National Academy of Sciences, 2001, 98 (21): 11967-11972.

[220] Cheung D T, DiCesare P, Benya P D, et al. The presence of intermolecular disulfide cross-links in type Ⅲ collagen. Journal of Biological Chemistry, 1983, 258 (12): 7774-7778.

[221] Zhang Y, Heher P, Hilborn J, et al. Hyaluronic acid-fibrin interpenetrating double network hydrogel prepared in situ by orthogonal disulfide cross-linking reaction for biomedical applications. Acta biomaterialia, 2016, 38: 23-32.

[222] Shu X Z, Liu Y, Luo Y, et al. Disulfide cross-linked hyaluronan hydrogels. Biomacromolecules, 2002, 3 (6): 1304-1311.

[223] Lee H, Park T G. Reduction/oxidation induced cleavable/crosslinkable temperature-sensitive hydrogel network containing disulfide linkages. Polymer journal, 1998, 30 (12): 976-980.

[224] Meng F, Hennink W E, Zhong Z. Reduction-sensitive polymers and bioconjugates for biomedical applications. Biomaterials, 2009, 30 (12): 2180-2198.

[225] Chen X, Lai N C H, Wei K, et al. Biomimetic presentation of cryptic ligands via single-chain nanogels for synergistic regulation of stem cells. ACS nano, 2020, 14 (4): 4027-4035.

[226] Lei Z Q, Xiang H P, Yuan Y J, et al. Room-temperature self-healable and remoldable cross-linked polymer based on the dynamic exchange of disulfide bonds. Chemistry of Materials, 2014, 26 (6): 2038-2046.

[227] Ryu J H, Chacko R T, Jiwpanich S, et al. Self-cross-linked polymer nanogels: a versatile nanoscopic drug delivery platform. Journal of the American Chemical Society, 2010, 132 (48): 17227-17235.

[228] Choh S Y, Cross D, Wang C. Facile synthesis and characterization of disulfide-cross-linked hyaluronic acid hydrogels for protein delivery and cell encapsulation. Biomacromolecules, 2011, 12 (4): 1126-1136.

[229] Wu D C, Loh X J, Wu Y L, et al. 'Living' controlled in situ gelling systems: thiol-disulfide

exchange method toward tailor-made biodegradable hydrogels. Journal of the American Chemical Society, 2010, 132 (43): 15140-15143.

[230] Burns J A, Butler J C, Moran J, et al. Selective reduction of disulfides by tris (2-carboxyethyl) phosphine. The Journal of Organic Chemistry, 1991, 56 (8): 2648-2650.

[231] Konigsberg W. Reduction of disulfide bonds in proteins with dithiothreitol. In Methods in enzymology, Elsevier, 1972, 25: 185-188.

[232] Pleasants J C, Guo W. Rabenstein, D. L., A comparative study of the kinetics of selenol/diselenide and thiol/disulfide exchange reactions. Journal of the American Chemical Society, 1989, 111 (17): 6553-6558.

[233] Deng G, Li F, Yu H, et al. Dynamic hydrogels with an environmental adaptive self-healing ability and dual responsive sol-gel transitions. ACS Macro Letters, 2012, 1 (2): 275-279.

[234] Deng G, Tang C, Li F, et al. Covalent cross-linked polymer gels with reversible sol- gel transition and self-healing properties. Macromolecules, 2010, 43 (3): 1191-1194.

[235] Yang X, Liu G, Peng L, et al. Highly efficient self-healable and dual responsive cellulose-based hydrogels for controlled release and 3D cell culture. Advanced Functional Materials, 2017, 27 (40): 1703174.

[236] Dirksen A, Dirksen S, Hackeng T M, et al. Nucleophilic catalysis of hydrazone formation and transimination: implications for dynamic covalent chemistry. Journal of the American Chemical Society, 2006, 128 (49): 15602-15603.

[237] Dirksen A, Yegneswaran S, Dawson P E. Bisaryl hydrazones as exchangeable biocompatible linkers. Angewandte Chemie, 2010, 122 (11): 2067-2071.

[238] Skene W G, Lehn J M P. Dynamers: polyacylhydrazone reversible covalent polymers, component exchange, and constitutional diversity. Proceedings of the National Academy of Sciences, 2004, 101 (22): 8270-8275.

[239] Liu F, Li F, Deng G, et al. Rheological images of dynamic covalent polymer networks and mechanisms behind mechanical and self-healing properties. Macromolecules, 2012, 45 (3): 1636-1645.

[240] Kool E T, Park D H, Crisalli P, Fast hydrazone reactants: electronic and acid/base effects strongly influence rate at biological pH. Journal of the American Chemical Society, 2013, 135 (47): 17663-17666.

[241] Nguyen R, Huc I. Optimizing the reversibility of hydrazone formation for dynamic combinatorial chemistry. Chemical communications, 2003, (8): 942-943.

[242] Dahlmann J, Krause A, Möller L, et al. Fully defined in situ cross-linkable alginate and hyaluronic acid hydrogels for myocardial tissue engineering. Biomaterials, 2013, 34 (4): 940-951.

[243] Patenaude M, Hoare T, Injectable, mixed natural-synthetic polymer hydrogels with modular properties. Biomacromolecules, 2012, 13 (2): 369-378.

[244] McKinnon D, Domaille D, Brown T, et al. Measuring cellular forces using bis-aliphatic hydrazone crosslinked stress-relaxing hydrogels. Soft matter, 2014, 10 (46): 9230-9236.

[245] Jung H, Park J S, Yeom J, et al. 3D tissue engineered supramolecular hydrogels for controlled chondrogenesis of human mesenchymal stem cells. Biomacromolecules, 2014, 15 (3): 707-714.

[246] Mukherjee S, Hill M R, Sumerlin B S. Self-healing hydrogels containing reversible oxime crosslinks. Soft matter, 2015, 11 (30): 6152-6161.

[247] Sánchez-Morán H, Ahmadi A, Vogler B, et al. Oxime cross-linked alginate hydrogels with tunable stress relaxation. Biomacromolecules, 2019, 20 (12): 4419-4429.

[248] Buffa R, Šedová P, Basarabová I, et al. α, β-Unsaturated aldehyde of hyaluronan—Synthesis, analysis and applications. Carbohydrate polymers, 2015, 134, 293-299.

[249] Grover G N, Lam J, Nguyen T H, et al. Biocompatible hydrogels by oxime click chemistry. Biomacromolecules, 2012, 13 (10): 3013-3017.

[250] Grover G N, Braden R L, Christman K L. Oxime cross-linked injectable hydrogels for catheter delivery. Advanced Materials, 2013, 25 (21): 2937-2942.

[251] Kalia J, Raines R T, Hydrolytic stability of hydrazones and oximes. Angewandte Chemie International Edition, 2008, 47 (39): 7523-7526.

[252] Lin F, Yu J, Tang W, et al. Peptide-functionalized oxime hydrogels with tunable mechanical properties and gelation behavior. Biomacromolecules, 2013, 14 (10): 3749-3758.

[253] Kloetzel M C. The Diels-Alder reactions with maleic anhydride. Org React, 1948, 4: 1-59.

[254] Holmes, H. The Diels-Alder Reaction Ethylenic and Acetylenic Dienophiles. Organic Reactions, 2004, 4: 60-173.

[255] Basilevskii M, Shamov A, Tikhomirov V. Transition state of the Diels-Alder reaction. Journal of the American Chemical Society, 1977, 99 (5): 1369-1372.

[256] Liu S, Lei Y, Qi X, Lan, Y. Reactivity for the Diels-Alder reaction of cumulenes: a distortion-interaction analysis along the reaction pathway. The Journal of Physical Chemistry A, 2014, 118 (14): 2638-2645.

[257] Nimmo C M, Owen S C, Shoichet M S. Diels-Alder click cross-linked hyaluronic acid hydrogels for tissue engineering. Biomacromolecules, 2011, 12 (3): 824-830.

[258] Shao C, Wang M, Chang H, et al. A self-healing cellulose nanocrystal-poly (ethylene glycol) nanocomposite hydrogel via Diels-Alder click reaction. ACS Sustainable Chemistry & Engineering, 2017, 5 (7): 6167-6174.

[259] Tan H, Rubin J P, Marra K G. Direct Synthesis of Biodegradable Polysaccharide Derivative Hydrogels through Aqueous Diels-Alder Chemistry. Macromolecular Rapid Communications, 2011, 32 (12): 905-911.

[260] Wei H L, Yang Z, Chu H J, et al. Facile preparation of poly (N-isopropylacrylamide)-based hydrogels via aqueous Diels-Alder click reaction. Polymer, 2010, 51 (8): 1694-1702.

[261] Kirchhof S, Brandl F P, Hammer N, et al. Investigation of the Diels-Alder reaction as a cross-linking mechanism for degradable poly (ethylene glycol) based hydrogels. Journal of Materials Chemistry B, 2013, 1 (37): 4855-4864.

[262] Adzima B J, Kloxin C J, Bowman C N. Externally triggered healing of a thermoreversible covalent network via self-limited hysteresis heating. Advanced Materials, 2010, 22 (25): 2784-2787.

[263] Chen X, Dam M A, Ono K, et al. A thermally re-mendable cross-linked polymeric material. Science, 2002, 295 (5560): 1698-1702.

[264] Gacal B, Durmaz H, Tasdelen M, et al. Anthracene-maleimide-based Diels-Alder "click

chemistry" as a novel route to graft copolymers. Macromolecules, 2006, 39 (16): 5330-5336.
[265] Moses J E, Moorhouse A D. The growing applications of click chemistry. Chemical Society Reviews, 2007, 36 (8): 1249-1262.
[266] Nandivada H, Jiang X, Lahann J. Click chemistry: versatility and control in the hands of materials scientists. Advanced Materials, 2007, 19 (17): 2197-2208.
[267] Wei Z, Yang J H, Du X J, et al. Dextran-based self-healing hydrogels formed by reversible Diels-Alder reaction under physiological conditions. Macromolecular rapid communications, 2013, 34 (18): 1464-1470.
[268] Koehler K C, Anseth K S, Bowman C N. Diels-Alder mediated controlled release from a poly (ethylene glycol) based hydrogel. Biomacromolecules, 2013, 14 (2): 538-547.

第 5 章
水凝胶高分子网络的结构特征

5.1 弹性体水凝胶
5.2 非弹性体水凝胶
5.3 由非常规高分子网络结构引起的力学性能分离
5.4 非常规高分子网络结构和相互作用的协同效应
参考文献

在本章中，我们将展开介绍传统高分子网络和非常规高分子网络，前者是我们通常用于计算高分子材料力学性能的简化模型，后者主要是生物组织中具有的各种不同类型的网络架构，它们能赋予水凝胶极端的力学性能。

5.1 弹性体水凝胶

传统高分子网络为通过永久共价键交联成网的高分子链，其中高分子链的缠结、物理交联和可逆交联可忽略不计[1-3]。传统高分子网络为非缠结弹性橡胶的发展提供了基本模型，包括仿射网络模型和幻影网络模型[1-3]。

5.1.1 干燥状态下的弹性高分子网络

在干燥状态下（图5.1），传统高分子网络中每单位体积包含n条高分子链，其中高分子链为两条连续共价交联之间的高分子链段。每条高分子链含有N个库恩单体（源于库恩柔性链模型，代指长度均一化的链段），每个库恩单体的长度为b。处于松弛状态和完全拉伸状态的高分子链的端到端距离分别为$\sqrt{N}b$ 和 Nb。因此，该高分子网络中高分子链的拉伸极限可根据式（5.1）计算[1-3]：

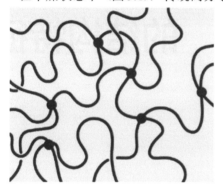

图5.1 干燥状态下常规高分子网络的示意图

$$\lambda_{\lim} = \frac{Nb}{\sqrt{N}b} = N^{1/2} \tag{5.1}$$

本体高分子网络的拉伸极限与链拉伸极限$\lambda_{\lim}$成正比，缩放关系的预因子取决于高分子的网络结构[4]。

假设干燥常规高分子网络遵循仿射网络模型，网络在初始变形下的剪切模量可表示为[1-3]：

$$G = nkT \tag{5.2}$$

式中，k是玻尔兹曼常数；T是热力学温度。

根据 Lake–Thomas 模型[5]，干燥高分子网络的断裂韧性是其固有断裂能 Γ_0，即每单位面积断裂单层高分子链所需的能量。

$$\Gamma_0 = n\sqrt{N}bNU_f = nbN^{3/2}U_f \qquad (5.3)$$

式中，$\sqrt{N}b$ 是每单位面积的高分子链数；NU_f 是使高分子链断裂所需的能量；U_f 是使单个库恩单体断裂所需的能量。

同样基于 Lake-Thomas 模型[5,6]，干燥高分子网络的疲劳阈值是固有断裂能 Γ_0。如果干燥常规高分子网络共价键合在基材上（图5.2），则界面韧性和界面疲劳阈值也都在 Γ_0 水平上[7-9]。

将 b、N、n、kT 和 U_f 代入式（5.1）～式（5.3），我们可以估计剪切模量 G 约在千帕到兆帕的数量级，链拉伸极限 λ_{lim} 可达几十（没有纠缠），固有断裂能 Γ_0 为每平方米几百焦耳[3]。

干燥高分子网络的力学性能相互耦合。通常假设干燥状态下高分子链占据高分子网络的主要体积，因此高分子网络的体积守恒如下：

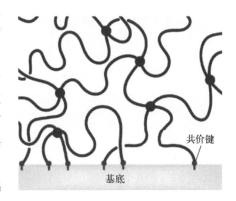

图5.2 干燥状态下常规高分子网络共价键合在基材上的示意图

$$Nnv = 1 \qquad (5.4)$$

式中，v 是库恩单体的体积。

通过将式（5.4）代入式（5.1）～式（5.3），我们可以将干燥状态下常规高分子网络的链拉伸极限 λ_{lim}、剪切模量 G 和固有断裂能 Γ_0 表示为其链长 N 的函数，如式（5.5）所示：

$$\lambda_{lim} = N^{1/2},\ G = N^{-1}v^{-1}kT,\ \Gamma_0 = N^{1/2}v^{-1}bU_f \qquad (5.5)$$

从式（5.5）可以明显看出，增加链长 N，能够增加链拉伸极限 λ_{lim} 和固有断裂能 Γ_0，但会降低常规高分子网络在干燥状态下的剪切模量 G。干燥状态下常规高分子网络的这些力学性能可以通过以下关系耦合：

$$\lambda_{lim} \sim \Gamma_0 \sim G^{-1/2} \qquad (5.6)$$

5.1.2 溶胀状态下的弹性高分子网络

5.1.1 中讨论了干燥常规高分子网络的模型参数,根据该模型,可预测干燥常规高分子网络可以吸水并膨胀成由常规高分子网络和水组成的水凝胶(图 5.3)。干燥高分子网络的溶胀以 λ_s 的比率拉伸网络中的高分子链,称为溶胀链拉伸。

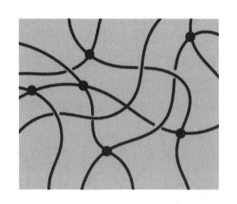

图 5.3 溶胀状态下常规高分子网络的示意图

由于干燥高分子网络的溶胀以 λ_s 的比率拉伸其高分子链,因此水凝胶中高分子链分别在松弛和完全拉伸状态下,端到端距离为 $\lambda_s\sqrt{Nb}$ 和 Nb。因此,水凝胶中高分子链的拉伸极限 λ_{lim} 可以表示为:

$$\lambda_{\text{lim}} = \frac{Nb}{\lambda_s\sqrt{Nb}} = N^{1/2}\lambda_s^{-1} \tag{5.7}$$

块状水凝胶的拉伸极限与链拉伸极限 λ_{lim} 成正比,缩放关系的预因子取决于高分子网络结构[4]。

干燥高分子网络的溶胀使其剪切模量降低了 λ_s 的比率[1]。因此,初始变形下水凝胶的剪切模量可表示为:

$$G = nkT\lambda_s^{-1} \tag{5.8}$$

需要注意的是,式(5.8)中的 n 是干燥高分子网络每单位体积的高分子链数。

干燥高分子网络的溶胀使得每单位面积的高分子链数减少了 λ_s^2 的比率,但这不会显著改变网络中高分子链断裂所需的能量。因此,水凝胶的固有断裂能 Γ_0 可以计算为:

$$\Gamma_0 = \frac{n\sqrt{Nb}}{\lambda_s^2}NU_f = nbN^{3/2}U_f\lambda_s^{-2} \tag{5.9}$$

具有传统高分子网络的水凝胶的断裂韧性和疲劳阈值是水凝胶的固有断裂能 Γ_0[5,6]。如果水凝胶的高分子网络是共价键合在基底上的(图 5.4),那么相应的界面韧性和界面疲劳阈值也都与水凝胶的固有断裂能 Γ_0 大致相同[7-9]。

通过比较式（5.1）～式（5.3）和式（5.7）～式（5.9）可以看出，将干燥高分子网络溶胀成水凝胶中降低干网络的链拉伸极限 λ_{lim}、剪切模量 G 以及固有断裂能 Γ_0，其因子分别为 λ_s、λ_s 和 λ_s^2 [1-3]。通过将 λ_s、b、N、n、kT 和 U_f 的典型值代入式（5.7）～式（5.9），我们可以估算出具有传统高分子网络的水凝胶的剪切模量 G 在帕到兆帕的量级，链条拉伸极限 λ_{lim} 在几倍左右（无缠结），固有断裂能 Γ_0 可以达到几十焦耳每平方米。

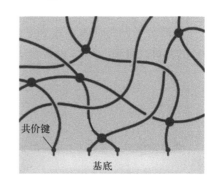

图 5.4 溶胀状态下常规高分子网络共价键合在基材上的示意图

通过将式（5.4）代入式（5.7）～式（5.9），我们可以同样将具有传统高分子网络的水凝胶的链拉伸极限 λ_{lim}、剪切模量 G 和固有断裂能 Γ_0 表示为其链长 N 的函数，如式（5.10）：

$$\lambda_{lim} = N^{1/2}\lambda_s^{-1}, G = N^{-1}v^{-1}kT\lambda_s^{-1}, \Gamma_0 = N^{1/2}v^{-1}bU_f\lambda_s^{-2} \quad (5.10)$$

从式（5.10）可以明显看出，增加链长 N 能够提升链拉伸极限 λ_{lim} 和固有断裂能 Γ_0，但会降低具有传统高分子网络的水凝胶的剪切模量 G。水凝胶的这些力学性能通过以下关系耦合：

$$\lambda_{lim} \sim \Gamma_0 \sim G^{-1/2} \quad (5.11)$$

值得注意的是，我们可以计算出由传统高分子网络的平衡溶胀而引起的链拉伸。在不失一般性的情况下，以具有立方体形状的干燥常规高分子网络为例。当高分子网络在水中达到平衡状态时，立方体的一侧将其长度从干燥状态增加一个比率 λ_{eq}。在平衡状态下，用于拉伸高分子链的每单位体积的干燥高分子网络的亥姆霍兹自由能 $W_{stretch}$，以及用于混合高分子和水的单位体积的干燥高分子网络亥姆霍兹自由能 W_{mix} 可表示为[1,10]：

$$W_{stretch} = \frac{1}{2}nkT\left(3\lambda_{eq}^2 - 3 - 6\lg\lambda_{eq}\right) \quad (5.12)$$

$$W_{mix} = -\frac{kT}{v_s}\left[\left(1-\lambda_{eq}^{-3}\right)\lg\left(1-\lambda_{eq}^{-3}\right) + \chi\lambda_{eq}^{-3}\right] \quad (5.13)$$

式中，χ 是 Flory 高分子-溶剂的相互作用参数；v_s 是溶剂分子的体积。因此，单位体积的干燥高分子网络的总亥姆霍兹自由能可表示为[1,10]：

$$W = W_{\text{stretch}} + W_{\text{mix}} \tag{5.14}$$

当高分子网络在水中达到平衡状态时，λ_{eq} 对应的总亥姆霍兹自由能是最低的[1,10]，可以得到：

$$\frac{\partial W}{\partial \lambda_{eq}} = 0 \tag{5.15}$$

通过求解式（5.15），可以得到水凝胶在平衡溶胀状态下的 λ_{eq}。一般而言，水凝胶中高分子链的拉伸 λ_s 与 λ_{eq} 成比例，而比例关系的预因子（比例因子）取决于高分子网络结构[4]。虽然式（5.12）～式（5.15）假设水凝胶的高分子网络是不带电的，但电荷对水凝胶溶胀平衡的影响可以通过在亥姆霍兹自由能函数中引入附加项来解释 [式（5.12）和式（5.13）][1,11]。需要注意的是，某些特定情形下，如当水凝胶与水绝缘或没有足够的时间与水平衡时，水凝胶不需要达到溶胀平衡状态。

5.2 非弹性体水凝胶

5.1 节中已经确定具有传统高分子网络的弹性体和水凝胶具有内在耦合的力学性能，包括剪切模量、拉伸极限、断裂韧性、疲劳阈值、界面黏附韧性和界面黏附疲劳阈值 [式（5.6）和式（5.11）]。本节将讨论非常规高分子网络（UPN），这是大多数生物水凝胶的网络结构，相应的设计机理已广泛用于合成人工水凝胶以实现极端的力学性能。

UPN 被定义为在网络结构和/或网络中高分子链之间的相互作用方式不同于传统高分子网络的高分子网络[12-21]。一般情况下，UPN 大致分为两类：UPN 架构和 UPN 交互。UPN 架构与由具有共价键的随机交联高分子链组成的传统高分子网络的架构具有显著不同。几乎所有生物组织（第 1 章）都具有特定类型的 UPN 架构。在过去的几十年里，科研人员已经为包括弹性体、水凝胶和有机凝胶在内的软材料设计和合成了多种 UPN 架构，以赋予这些软材料优异性能。根据其拓扑结构，UPN 架构可进一步分为理想高分子网络、具有可滑动交联的高分子网络、互穿高分子网络、半互穿高分子网络、具有高官能交联的高分子网络、微纳米纤维高分子网络和其他非弹性体高分子网络。

5.2.1 理想高分子网络

理想高分子网络是指链长均匀、功能均匀且无缺陷的高分子网络（图5.5）[22]。理想高分子网络通常使用多臂大分子单体制造，其中相邻大分子单体的臂交联成高分子链[22,23]。由于大分子臂长度均匀，且交联过程反应效率高，因此可以制备得到各种链长均匀、功能均匀、几乎无缺陷的理想高分子网络[22,24-35]。四臂PEG，作为最常用的制造具有理想高分子网络的水凝胶的大分子单体之一[22,36,37]，通常会用N-羟基琥珀酰亚胺和胺[22,24,38]、四苯甲醛和四苯甲酰肼[39]、马来酰亚胺和硫醇[40]、硼酸和二醇[25,29,36]等成对的反应基团对其末端进行修饰。由于几乎没有缺陷，理想高分子网络具有高度可拉伸和高弹性等性质[24]。需要注意的是，虽然传统的高分子网络通

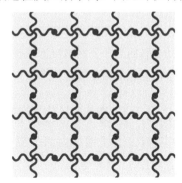

图5.5 理想高分子网络架构的示意图

常存在不均匀的链长和拓扑缺陷，但它们的力学性能通常是基于理想高分子网络的模型计算的，因此，理想高分子网络本身仍然具有耦合的力学性能（指代该分子网络中的各分子链之间的力学性能对整体的性能存在紧密配合和相互影响）。

5.2.2 含有滑动交联点的高分子网络

可滑动的交联点，通常以两个共价交联的高分子环的形式存在，可以将两条穿过环并在环内滑动的高分子链相互连接（图5.6）[41]。由于交联的永久性和可滑动性，具有可滑动交联点的高分子网络兼具机械稳定性和可重构性。在机械载荷下，

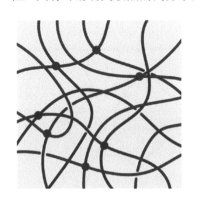

图5.6 具有可滑动交联点的高分子网络架构的示意图

可滑动的交联点倾向于重新配置高分子网络，使网络中的高分子链承受相同水平的力，因此重新配置的高分子网络接近理想高分子网络。

具有可滑动交联点的高分子网络主要由基于环糊精的聚轮烷合成[42-44]。基于环糊精的聚轮烷是由线型高分子链组成的包合物，线型高分子链穿过环糊精分子，然后在链端被庞大的基团封端[44-46]。基于环糊精的聚轮烷的形成条件主要取决于环糊精内部空腔与高分子链横截面之间的尺寸匹配度[47]。研究人员已经成功使用不同的高分子链构建出不同类型的基于环糊

精的聚轮烷，包括线型均聚物、线型嵌段共聚物以及支化高分子[47]。其中，α-环糊精具有最小的空腔尺寸，可以与 PEG 或 PCL 形成包合物，但不能与聚对苯醚（PPO）链形成包合物[48,49]。β-环糊精可以与 PCL 或 PPO 形成包合物，但不能与 PEG 形成包合物[48,50,51]。具有最大空腔尺寸的 γ-环糊精内部可以穿过一条 PPO 链或两条 PEG 或 PCL 链[52]。此外，环糊精之间可以相互交联，从而使得线型高分子链相互连接，并形成具有可滑动交联的高分子网络[53,54]。由于在机械载荷下具有可滑动交联的高分子网络接近理想高分子网络，因此具有可滑动交联的高分子网络的力学性能通常相互耦合[41,55-57]。

5.2.3　互穿和半互穿高分子网络

互穿高分子网络由两个或多个小的互穿高分子网络组成，它们单独交联但没有连接在一起（图 5.7）；半互穿高分子网络同样是由两个或多个小的互穿高分子网络组成，只是其中至少有一个网络未交联，其他网络单独交联但未连接在一起（图 5.8）[54, 58-66]。互穿和半互穿高分子网络相互缠绕或互锁，除非网络被破坏，否则它们将一直保持互穿结构[63-66]。基于互穿和半互穿高分子网络的水凝胶通常按照"顺序法"或"同步法"制备。在顺序法中，将一种高分子网络浸入另一种高分子网络的单体、引发剂和/或交联剂的溶液中，通过在已有的网络内聚合新的高分子网络来形成互穿或半互穿高分子网络[12]。在同步法中，所有高分子网络的高分子、单体、引发剂和交联剂的混合物一步法或一锅法形成互穿或半互穿高分子网络[67,68]。与顺序法相比，这种一步法或一锅法制造工艺是同步法的一个理想特征。包括第 2 章中讨论的天然高分子和合成高分子在内的各种候选材料已被用于通过各种交联策略合成具有互穿和半互穿高分子网络的水凝胶[58,62,69,70]。正如之前讨论的那样，互穿和半互穿高分子网络可以为水凝胶提供解耦和极端的力学性能，例如极高的拉伸性和断裂韧性[12,68,71-73]。

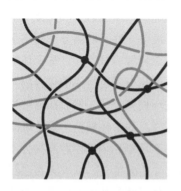 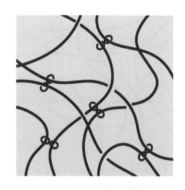

图 5.7　互穿高分子网络架构的示意图　图 5.8　半互穿高分子网络架构的示意图

5.2.4 具有高官能交联的高分子网络

高官能交联高分子网络中的交联官能是指在交联处相互连接的高分子链的数量。第 4 章中讨论的常见共价交联通常具有相对较低的官能度（通常小于 10），并且通常在两个相邻的共价交联之间存在单个高分子链桥接。为了显著增强高分子网络的功能，可以在高分子网络中引入各种类型的高官能交联，包括结晶域[74-77]、玻璃状结节[78,79]、微纳米颗粒[54,58,77,80-83]、微相分离[84-86]等（图 5.9）。例如，聚乙烯醇可以通过冻融法形成纳米晶域以交联高分子网络[76,87]，聚甲基丙烯酸甲酯可以形成玻璃球体，将基于聚甲基丙烯酸甲酯的嵌段共聚物交联成网络[88]；纳米黏土[89]和氧化石墨烯[90]等剥离的颗粒，可以将聚丙烯酰胺交联成可模塑和可自修复的水凝胶；苯乙烯、丙烯酸丁酯和丙烯酸的混合物可以形成微球，将残留的高分子链交联成微球复合水凝胶[91]。

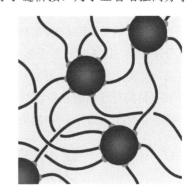

图 5.9 具有高官能交联的高分子网络架构的示意图

多个高分子链（通常超过 10 个）可以在每个高官能交联处相互连接（图 5.9）。此外，两个相邻的高官能交联之间可能存在多个高分子链桥接，其中高分子链的长度可能非常不均匀[26,53]。正如之前讨论的那样，具有高官能交联的高分子网络可以为水凝胶提供解耦和高断裂韧性、弹性、拉伸强度和抗疲劳性等高力学性能。

5.2.5 微纳纤维高分子网络

合成高分子和天然高分子均可通过共价键或物理键组装成微纳级直径的纤维（或原纤维，指的是短纤维），并进一步通过缠结、聚合和交联成渗透高分子网络（图 5.10）[92-99]。在生物体中，细胞可以分泌蛋白质（胶原蛋白）和多糖（纤维素），然后组装成微纳纤维高分子网络[100-105]。这些天然衍生的纤维及其网络可用于制备具有微纳纤维高分子网络的水凝胶[106-110]。此外，各种天然高分子和合成高分子还可以通过纺丝技术制成微纳纤维高分子网络[111-113]。在诸多纺丝技术中，静电纺丝技术在操作中只需通过调整静电纺丝过程的参数，就可以轻易控

图 5.10 微纳纤维高分子网络架构的示意图

制纤维的直径、排列和密度[115-119]，因而具有工艺简单、成本低廉和广泛的适用性等特点，从而受到广泛关注[114]。微纳纤维高分子网络同样可以为水凝胶提供解耦和极端的力学性能[120,121]。

5.2.6 其他非常规高分子网络

许多其他类型的UPN架构因各自独特的性能而受到科研人员的关注。如刷状高分子网络（图5.11）在无溶剂状态下可以显示出极低的剪切模量和类似组织的应力-应变关系[122,123]，尽管这一类UPN架构尚未广泛用于水凝胶，但具有应用于水凝胶的设计的潜在价值[124]。随着高分子和软材料的进一步开发，新的UPN架构及非弹性体高分子网络体系也将随之出现。

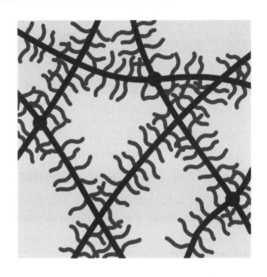

图5.11　刷状高分子网络架构的示意图

5.3　由非常规高分子网络结构引起的力学性能分离

理想高分子网络中的高分子链具有均匀的链长（即相同的 N），具有可滑动交联点的高分子网络在机械载荷下也倾向于提供相对均匀的链长。这些具有相对均匀链长的高分子网络称为"单峰高分子网络"（图5.12）[18,125,126]。因为常规高分子网络的剪切模量、拉伸极限和固有断裂能是基于单峰高分子网络推导出来的，所以这些力学性能在具有理想高分子网络和具有可滑动交联点的高分子网络水凝胶中仍然耦合。

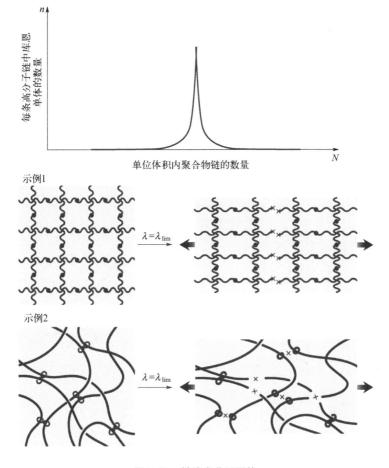

图 5.12 单峰高分子网络

单峰高分子网络,如理想高分子网络和具有可滑动交联点的高分子网络,具有耦合的力学性能

互穿高分子网络、半互穿高分子网络和具有高官能交联的高分子网络可以将具有不同链长(即不同 N)的高分子链整合到相同的高分子网络中,这些网络通常被称为多峰高分子网络(图 5.13)[18,125,126]。在此,我们根据链长将多峰高分子网络中的高分子链分为以下类型:对于第 i 类高分子链,干燥状态下单位体积高分子网络中的高分子链数、每条高分子链的库恩单体数以及库恩单体的体积分别表示为 n_i、N_i 和 v_i。

因此,可以设计多峰高分子网络(图 5.13),使相应的水凝胶可以在最长高分子链的拉伸极限内保持其完整性,可以表示为[12,68]:

$$\lambda_{\lim} = \sqrt{N_{\max}}\lambda_s^{-1} \tag{5.16}$$

式中，N_{max} 是最长高分子链上库恩单体的数量；λ_s^{-1} 是溶胀对链拉伸极限 λ_{lim} 的影响。块状水凝胶的拉伸极限和缩放关系的预因子取决于高分子网络结构[4]。

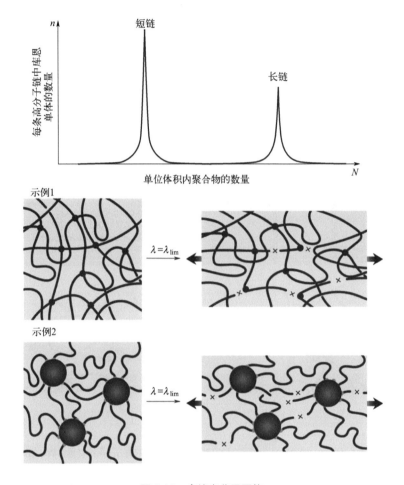

图 5.13　多峰高分子网络

多峰高分子网络，如互穿高分子网络、半互穿高分子网络和具有高官能交联的
高分子网络，可以解耦力学性能

在仿射网络模型的基础上，具有多峰高分子网络（图 5.13）的水凝胶的剪切模量可表示为：

$$G = \sum n_i kT \lambda_s^{-1} \tag{5.17}$$

式中，n_i、λ_s^{-1} 解释了第 i 类高分子链和溶胀对水凝胶初始剪切模量的影响。

按照 Lake–Thomas 模型，具有多峰高分子网络的水凝胶的固有断裂能可以表示为：

$$\varGamma_0 = \sum n_i b_i N_i^{3/2} U_i \lambda_s^{-2} \qquad (5.18)$$

式中，b_i、U_i 分别是第 i 类高分子链上库恩单体的长度和断裂能；λ_s^{-2} 是溶胀对固有断裂能的影响。需要注意的是，多峰高分子网络的断裂韧性和界面断裂韧性可能远高于固有断裂能，这将在下一章中讨论。

通常假设高分子链在干燥状态下占据高分子网络的主要体积，因此依据多峰高分子网络的体积守恒可以得出：

$$\sum N_i n_i v_i = 1 \qquad (5.19)$$

尽管存在式（5.19）的关系，但具有多峰高分子网络的水凝胶的拉伸极限、剪切模量和固有断裂能仍然可以解耦和独立设计。例如，以链长是双峰分布的水凝胶为例，短高分子链的高密度可以赋予水凝胶高的初始剪切模量。虽然当水凝胶被高度拉伸时这些短链会断裂，但高分子中的长链仍然可以维持水凝胶的完整性和高拉伸极限[12,68]。因此，长高分子链可以使水凝胶具有相对较高的固有断裂能[127]。

微纳纤维高分子网络的力学性能由纤维材料的性质、纤维间的相互作用（如纤维间的交联）以及纤维高分子网络的拓扑结构共同决定。因此，微纳纤维水凝胶的拉伸极限、剪切模量和固有断裂能不遵循传统高分子网络的耦合关系[式（5.14）和式（5.15）]，所以它们可以独立设计。

5.4 非常规高分子网络结构和相互作用的协同效应

非常规高分子网络同时拥有 UPN 架构和 UPN 交互的情况并不少见。在某些情况下，UPN 体系结构和交互是相互依赖的。例如，某些 UPN 架构的形成需要某些 UPN 相互作用的存在，抑或是某些 UPN 相互作用会导致高分子自组装到某些 UPN 架构中。一个典型的例子是结晶域和玻璃结节等强物理交联，它们的 UPN 相互作用大多具有高官能，因而会产生具有高官能交联的 UPN 架构。此外，高分子链自组装成微纳纤维通常也需要弱物理交联等 UPN 相互作用。

一些特定类型的 UPN 架构和 UPN 交互可以单独设计，然后集成到同一个 UPN 中。例如，弱物理交联和动态共价交联已被引入各种 UPN 结构中以设计高韧性水凝胶[68]，因为这些可逆交联的解离和重组可以耗散机械能以增韧水凝胶。

参考文献

[1] Flory P J. Principles of polymer chemistry. Cornell university press,1953.

[2] Treloar L G.The physics of rubber elasticity. 1975.

[3] Rubinstein M,Colby R H.Polymer physics. Oxford university press New York:2003,23.

[4] Boyce M C,Arruda E M. Constitutive models of rubber elasticity: a review. Rubber chemistry and technology,2000,73 (3): 504-523.

[5] Lake G,Thomas A. The strength of highly elastic materials. Proceedings of the Royal Society of London Series A Mathematical and Physical Sciences,1967,300 (1460): 108-119.

[6] Lake G,Lindley P. The mechanical fatigue limit for rubber. Journal of Applied Polymer Science,1965,9 (4): 1233-1251.

[7] Yuk H,Zhang T,Lin S,et al. Tough bonding of hydrogels to diverse non-porous surfaces. Nature materials,2016,15 (2): 190-196.

[8] Liu J,Lin S,Liu X,et al, Fatigue-resistant adhesion of hydrogels. Nature communications,2020,11 (1): 1071.

[9] Ahagon A,Gent A. Effect of interfacial bonding on the strength of adhesion. Journal of Polymer Science: Polymer Physics Edition 1975,13 (7): 1285-1300.

[10] Hong W,Zhao X,Zhou J,et al. A theory of coupled diffusion and large deformation in polymeric gels. Journal of the Mechanics and Physics of Solids,2008,56 (5): 1779-1793.

[11] Hong W,Zhao X,Suo Z. Large deformation and electrochemistry of polyelectrolyte gels. Journal of the Mechanics and Physics of Solids,2010,58 (4): 558-577.

[12] Gong J P,Katsuyama Y,Kurokawa T,et al. Double-network hydrogels with extremely high mechanical strength. Advanced materials,2003,15 (14): 1155-1158.

[13] Roland C. Unconventional rubber networks: Circumventing the compromise between stiffness and strength. Rubber Chemistry and Technology,2013,86 (3): 351-366.

[14] Richtering W,Saunders B R. Gel architectures and their complexity. Soft Matter,2014,10 (21): 3695-3702.

[15] Wang H,Heilshorn S C. Adaptable hydrogel networks with reversible linkages for tissue engineering. Advanced Materials,2015,27 (25): 3717-3736.

[16] Petka W A,Harden J L,McGrath K P,et al. Reversible hydrogels from self-assembling artificial proteins. Science,1998,281 (5375): 389-392.

[17] Rosales A M,Anseth K S. The design of reversible hydrogels to capture extracellular matrix dynamics. Nature Reviews Materials,2016,1 (2): 1-15.

[18] Mark J E. Elastomeric networks with bimodal chain-length distributions. Accounts of chemical research,1994,27 (9): 271-278.

[19] Kloxin A M,Kasko A M,Salinas C N,et al. Photodegradable hydrogels for dynamic tuning of physical and chemical properties. Science,2009,324 (5923): 59-63.

[20] Creton C. 50th anniversary perspective: Networks and gels: Soft but dynamic and tough. Macromolecules,2017,50 (21): 8297-8316.

[21] Liu Y,He W,Zhang Z,et al. Recent developments in tough hydrogels for biomedical applications.

Gels, 2018, 4 (2): 46.

[22] Sakai T, Matsunaga T, Yamamoto Y, et al. Design and fabrication of a high-strength hydrogel with ideally homogeneous network structure from tetrahedron-like macromonomers. Macromolecules, 2008, 41 (14): 5379-5384.

[23] Sakai T. Physics of polymer gels. John Wiley & Sons, 2020.

[24] Kamata H, Akagi Y, Kayasuga-Kariya Y, et al. "Nonswellable" hydrogel without mechanical hysteresis. Science, 2014, 343 (6173): 873-875.

[25] Parada G A, Zhao X. Ideal reversible polymer networks. Soft Matter, 2018, 14 (25): 5186-5196.

[26] Akagi Y, Matsunaga T, Shibayama M, et al. Evaluation of topological defects in tetra-PEG gels. Macromolecules, 2010, 43 (1): 488-493.

[27] Yigit S, Sanyal R, Sanyal A. Fabrication and functionalization of hydrogels through "click" chemistry. Chemistry-An Asian Journal, 2011, 6 (10): 2648-2659.

[28] Huang X, Nakagawa S, Li X, et al. A simple and versatile method for the construction of nearly ideal polymer networks. Angewandte Chemie International Edition, 2020, 59 (24): 9646-9652.

[29] Marco-Dufort B, Iten R, Tibbitt M W. Linking molecular behavior to macroscopic properties in ideal dynamic covalent networks. Journal of the American Chemical Society, 2020, 142 (36): 15371-15385.

[30] Matsunaga T, Sakai T, Akagi Y, et al. SANS and SLS studies on tetra-arm PEG gels in as-prepared and swollen states. Macromolecules, 2009, 42 (16): 6245-6252.

[31] Hild G. Model networks based on endlinking' processes: synthesis, structure and properties. Progress in polymer science, 1998, 23 (6): 1019-1149.

[32] Zhong M, Wang R, Kawamoto K, et al. Quantifying the impact of molecular defects on polymer network elasticity. Science, 2016, 353 (6305): 1264-1268.

[33] Okaya Y, Jochi Y, Seki T, et al. Precise synthesis of a homogeneous thermoresponsive polymer network composed of four-branched star polymers with a narrow molecular weight distribution. Macromolecules, 2019, 53 (1): 374-386.

[34] Apostolides D E, Patrickios C S, Sakai T, et al. Near-model amphiphilic polymer conetworks based on four-arm stars of poly (vinylidene fluoride) and poly (ethylene glycol): synthesis and characterization. Macromolecules, 2018, 51 (7): 2476-2488.

[35] Oshima K, Fujimoto T, Minami E, et al. Model polyelectrolyte gels synthesized by end-linking of tetra-arm polymers with click chemistry: synthesis and mechanical properties. Macromolecules, 2014, 47 (21): 7573-7580.

[36] Yesilyurt V, Webber M J, Appel E A, et al. Injectable self-healing glucose-responsive hydrogels with pH-regulated mechanical properties. Advanced materials, 2016, 28 (1): 86-91.

[37] Shibayama M, Li X, Sakai T. Precision polymer network science with tetra-PEG gels—a decade history and future. Colloid and Polymer Science, 2019, 297: 1-12.

[38] Hashimoto K, Fujii K, Nishi K, et al. Nearly ideal polymer network ion gel prepared in pH-buffering ionic liquid. Macromolecules, 2016, 49 (1): 344-352.

[39] Apostolides D E, Sakai T, Patrickios C S. Dynamic covalent star poly (ethylene glycol) model hydrogels: A new platform for mechanically robust, multifunctional materials. Macromolecules,

2017, 50 (5): 2155-2164.

[40] Hayashi K, Okamoto F, Hoshi S, et al. Fast-forming hydrogel with ultralow polymeric content as an artificial vitreous body. Nature Biomedical Engineering, 2017, 1 (3): 44.

[41] Okumura Y, Ito K. The polyrotaxane gel: A topological gel by figure-of-eight cross-links. Advanced materials, 2001, 13 (7): 485-487.

[42] Wenz G, Han B H, Müller A. Cyclodextrin rotaxanes and polyrotaxanes. Chemical reviews, 2006, 106 (3): 782-817.

[43] Bin Imran A, Esaki K, Gotoh H, et al. Extremely stretchable thermosensitive hydrogels by introducing slide-ring polyrotaxane cross-linkers and ionic groups into the polymer network. Nature communications, 2014, 5 (1): 5124.

[44] Loethen S, Kim J M, Thompson D H. Biomedical applications of cyclodextrin based polyrotaxanes. Journal of Macromolecular Science, Part C: Polymer Reviews, 2007, 47 (3): 383-418.

[45] Mayumi K, Ito K. Structure and dynamics of polyrotaxane and slide-ring materials. Polymer, 2010, 51 (4): 959-967.

[46] Huang F, Gibson H W. Polypseudorotaxanes and polyrotaxanes. Progress in Polymer Science, 2005, 30 (10): 982-1018.

[47] Liu G, Yuan Q, Hollett G, et al.Cyclodextrin-based host–guest supramolecular hydrogel and its application in biomedical fields. Polymer Chemistry, 2018, 9 (25): 3436-3449.

[48] Harada A, Kamachi M. Complex formation between poly (ethylene glycol) and α-cyclodextrin. Macromolecules, 1990, 23 (10): 2821-2823.

[49] Harada A, Li J, Kamachi M. The molecular necklace: a rotaxane containing many threaded α-cyclodextrins. Nature, 1992, 356 (6367): 325-327.

[50] Xie D, Yang K, Sun W. Formation and characterization of polylactide and β-cyclodextrin inclusion complex. Current Applied Physics, 2007, 7: e15-e18.

[51] Harada A, Takashima Y, Yamaguchi H. Cyclodextrin-based supramolecular polymers. Chemical Society Reviews, 2009, 38 (4): 875-882.

[52] Gao P, Wang J, Ye L, et al. Stable and Unconventional Conformation of Single PEG Bent γ-CD-Based Polypseudorotaxanes. Macromolecular Chemistry and Physics, 2011, 212 (21): 2319-2327.

[53] Zhao X, Multi-scale multi-mechanism design of tough hydrogels: building dissipation into stretchy networks. Soft matter, 2014, 10 (5): 672-687.

[54] Peak C W, Wilker J J, Schmidt G. A review on tough and sticky hydrogels. Colloid and Polymer Science, 2013, 291: 2031-2047.

[55] Granick S, Rubinstein M. A multitude of macromolecules. Nature Materials, 2004, 3 (9): 586-587.

[56] Oku T, Furusho Y, Takata T. A Concept for Recyclable Cross-Linked Polymers: Topologically Networked Polyrotaxane Capable of Undergoing Reversible Assembly and Disassembly. Angewandte Chemie International Edition, 2004, 43 (8): 966-969.

[57] Ooya T, Eguchi M, Yui N. Enhanced accessibility of peptide substrate toward membrane-bound metalloexopeptidase by supramolecular structure of polyrotaxane. Biomacromolecules, 2001, 2 (1): 200-203.

[58] Wang W, Narain R, Zeng H. Rational design of self-healing tough hydrogels: a mini review. Frontiers

in chemistry, 2018, 6: 497.

[59] Li J, Suo Z, Vlassak J J. Stiff, strong, and tough hydrogels with good chemical stability. Journal of Materials Chemistry B, 2014, 2 (39): 6708-6713.

[60] Nonoyama T, Gong J P. Double-network hydrogel and its potential biomedical application: A review Proceedings of the Institution of Mechanical Engineers, Part H: Journal of Engineering in Medicine, 2015, 229 (12): 853-863.

[61] Costa A M, Mano J F. Extremely strong and tough hydrogels as prospective candidates for tissue repair-A review. European Polymer Journal, 2015, 72: 344-364.

[62] Dragan E S. Design and applications of interpenetrating polymer network hydrogels. A review. Chemical Engineering Journal, 2014, 243: 572-590.

[63] Sperling L H. Interpenetrating polymer networks and related materials. Springer Science & Business Media, 2012.

[64] Myung D, Waters D, Wiseman M, et al. Progress in the development of interpenetrating polymer network hydrogels. Polymers for advanced technologies, 2008, 19 (6): 647-657.

[65] Sperling L H, Hu R. Interpenetrating polymer networks. In Polymer blends handbook, Springer, 2014: 677-724.

[66] Visscher K B, Manners I, Allcock H R. Synthesis and properties of polyphosphazene interpenetrating polymer networks. Macromolecules, 1990, 23 (22): 4885-4886.

[67] Wang J J, Liu F. Enhanced adsorption of heavy metal ions onto simultaneous interpenetrating polymer network hydrogels synthesized by UV irradiation. Polymer bulletin, 2013, 70: 1415-1430.

[68] Sun J Y, Zhao X, Illeperuma W R, et al. Highly stretchable and tough hydrogels. Nature, 2012, 489 (7414): 133-136.

[69] Hennink W E, van Nostrum C F. Novel crosslinking methods to design hydrogels, 2012, 64: 223-236.

[70] Seiffert S, Kumacheva E, Okay O, et al. Supramolecular polymer networks and gels. Springer, 2015, 268.

[71] Tong X, Yang F. Engineering interpenetrating network hydrogels as biomimetic cell niche with independently tunable biochemical and mechanical properties. Biomaterials, 2014, 35 (6): 1807-1815.

[72] Feig V R, Tran H, Lee M, et al.Mechanically tunable conductive interpenetrating network hydrogels that mimic the elastic moduli of biological tissue. Nature communications, 2018, 9 (1): 2740.

[73] Darnell M C, Sun J Y, Mehta M, et al.Performance and biocompatibility of extremely tough alginate/polyacrylamide hydrogels. Biomaterials, 2013, 34 (33): 8042-8048.

[74] Croisier F, Jérôme C. Chitosan-based biomaterials for tissue engineering. European polymer journal, 2013, 49 (4): 780-792.

[75] Hassan C M, Peppas N A. Structure and morphology of freeze/thawed PVA hydrogels. Macromolecules, 2000, 33 (7): 2472-2479.

[76] Peppas N A, Merrill E W. Development of semicrystalline poly (vinyl alcohol) hydrogels for biomedical applications. Journal of biomedical materials research, 1977, 11 (3): 423-434.

[77] Kumar A, Han S S. PVA-based hydrogels for tissue engineering: A review. International journal of polymeric materials and polymeric biomaterials, 2017, 66 (4): 159-182.

[78] Henderson K J, Zhou T C, Otim K J, et al. Ionically cross-linked triblock copolymer hydrogels with high strength. Macromolecules, 2010, 43 (14): 6193-6201.

[79] Vancaeyzeele C, Fichet O, Boileau S, et al. Polyisobutene-poly (methylmethacrylate) interpenetrating polymer networks: synthesis and characterization. Polymer, 2005, 46 (18): 6888-6896.

[80] Schexnailder P, Schmidt G. Nanocomposite polymer hydrogels. Colloid and Polymer Science, 2009, 287: 1-11.

[81] Fu J. Strong and tough hydrogels crosslinked by multi-functional polymer colloids. Journal of Polymer Science Part B: Polymer Physics, 2018, 56 (19): 1336-1350.

[82] Liu Z, Faraj Y, Ju X J, et al. Nanocomposite smart hydrogels with improved responsiveness and mechanical properties: A mini review. Journal of Polymer Science Part B: Polymer Physics, 2018, 56 (19): 1306-1313.

[83] Hong S, Sycks D, Chan H F, et al. 3D printing of highly stretchable and tough hydrogels into complex, cellularized structures. Advanced materials, 2015, 27 (27): 4035-4040.

[84] Tuncaboylu D C, Sari M, Oppermann W, et al. Tough and self-healing hydrogels formed via hydrophobic interactions. Macromolecules, 2011, 44 (12): 4997-5005.

[85] Okay O. Self-healing hydrogels formed via hydrophobic interactions. Supramolecular Polymer Networks and Gels, 2015: 101-142.

[86] Jiang H, Duan L, Ren X, et al. Hydrophobic association hydrogels with excellent mechanical and self-healing properties. European Polymer Journal, 2019, 112: 660-669.

[87] Peppas N A, Merrill E W. Poly (vinyl alcohol) hydrogels: Reinforcement of radiation-crosslinked networks by crystallization. Journal of Polymer Science: Polymer Chemistry Edition, 1976, 14 (2): 441-457.

[88] Seitz M E, Martina D, Baumberger T, et al. Fracture and large strain behavior of self-assembled triblock copolymer gels. Soft Matter, 2009, 5 (2): 447-456.

[89] Okay O, Oppermann W. Polyacrylamide-clay nanocomposite hydrogels: rheological and light scattering characterization. Macromolecules, 2007, 40 (9): 3378-3387.

[90] Liu R, Liang S, Tang X Z, et al. Tough and highly stretchable graphene oxide/polyacrylamide nanocomposite hydrogels. Journal of Materials Chemistry, 2012, 22 (28): 14160-14167.

[91] Huang T, Xu H, Jiao K, et al. A novel hydrogel with high mechanical strength: a macromolecular microsphere composite hydrogel. Advanced Materials, 2007, 19 (12): 1622-1626.

[92] Moutos F T, Freed L E, Guilak F.A biomimetic three-dimensional woven composite scaffold for functional tissue engineering of cartilage. Nature materials, 2007, 6 (2): 162-167.

[93] Lee K Y, Mooney D J. Hydrogels for tissue engineering. Chemical reviews, 2001, 101 (7): 1869-1880.

[94] Abe K, Yano H. Formation of hydrogels from cellulose nanofibers. Carbohydrate Polymers, 2011, 85 (4): 733-737.

[95] Gao Y, Kuang Y, Guo Z F, et al. Enzyme-instructed molecular self-assembly confers nanofibers and a supramolecular hydrogel of taxol derivative. Journal of the American Chemical Society, 2009, 131 (38): 13576-13577.

[96] Ma M, Kuang Y, Gao Y, et al. Aromatic-aromatic interactions induce the self-assembly of pentapeptidic derivatives in water to form nanofibers and supramolecular hydrogels. Journal of the American Chemical Society, 2010, 132 (8): 2719-2728.

[97] Köhler K, Förster G, Hauser A, et al. Self-Assembly in a Bipolar Phosphocholine-Water System: The Formation of Nanofibers and Hydrogels. Angewandte Chemie International Edition, 2004, 43 (2): 245-247.

[98] Yang G, Lin H, Rothrauff B B, et al. Multilayered polycaprolactone/gelatin fiber-hydrogel composite for tendon tissue engineering. Acta biomaterialia, 2016, 35: 68-76.

[99] Ma L, Yang G, Wang N, et al. Trap Effect of Three-Dimensional Fibers Network for High Efficient Cancer-Cell Capture. Advanced healthcare materials, 2015, 4 (6): 838-843.

[100] Hoffman A S. Hydrogels for biomedical applications. Advanced drug delivery reviews, 2012, 64: 18-23.

[101] Moon R J, Martini A, Nairn, et al. Cellulose nanomaterials review: structure, properties and nanocomposites. Chemical Society Reviews, 2011, 40 (7): 3941-3994.

[102] D'Amore A, Stella J A, Wagner W R, et al. Characterization of the complete fiber network topology of planar fibrous tissues and scaffolds. Biomaterials, 2010, 31 (20): 5345-5354.

[103] Toshima M, Ohtani Y, Ohtani O. Three-dimensional architecture of elastin and collagen fiber networks in the human and rat lung. Archives of histology and cytology, 2004, 67 (1): 31-40.

[104] Weber K T. Cardiac interstitium in health and disease: the fibrillar collagen network. Journal of the American College of Cardiology, 1989, 13 (7): 1637-1652.

[105] Nakagaito, A, Iwamoto S, Yano H. Bacterial cellulose: the ultimate nano-scalar cellulose morphology for the production of high-strength composites. Applied Physics A, 2005, 80: 93-97.

[106] Ullah F, Othman M B H, Javed F, et al. Classification, processing and application of hydrogels: A review. Materials Science and Engineering: C, 2015, 57: 414-433.

[107] Yan C, Pochan D J. Rheological properties of peptide-based hydrogels for biomedical and other applications. Chemical Society Reviews, 2010, 39 (9): 3528-3540.

[108] Woolfson D N. Building fibrous biomaterials from α-helical and collagen-like coiled-coil peptides. Peptide Science: Original Research on Biomolecules, 2010, 94 (1): 118-127.

[109] Viebke C, Piculell L, Nilsson S. On the mechanism of gelation of helix-forming biopolymers. Macromolecules, 1994, 27 (15): 4160-4166.

[110] Prince E, Kumacheva E. Design and applications of man-made biomimetic fibrillar hydrogels. Nature Reviews Materials, 2019, 4 (2): 99-115.

[111] Tamayol A, Akbari M, Annabi N, et al. Fiber-based tissue engineering: Progress, challenges, and opportunities. Biotechnology advances, 2013, 31 (5): 669-687.

[112] Iwamoto S, Isogai A, Iwata T. Structure and mechanical properties of wet-spun fibers made from natural cellulose nanofibers. Biomacromolecules, 2011, 12 (3): 831-836.

[113] Ren L, Pandit V, Elkin J, et al. Large-scale and highly efficient synthesis of micro-and nano-fibers with controlled fiber morphology by centrifugal jet spinning for tissue regeneration. Nanoscale, 2013, 5 (6): 2337-2345.

[114] Agarwal S, Greiner A, Wendorff J H. Functional materials by electrospinning of polymers. Progress in

Polymer Science, 2013, 38 (6): 963-991.

[115] Rogina A. Electrospinning process: Versatile preparation method for biodegradable and natural polymers and biocomposite systems applied in tissue engineering and drug delivery. Applied Surface Science, 2014, 296: 221-230.

[116] Hong Y, Huber A, Takanari K, et al. Mechanical properties and in vivo behavior of a biodegradable synthetic polymer microfiber-extracellular matrix hydrogel biohybrid scaffold. Biomaterials, 2011, 32 (13): 3387-3394.

[117] Ekaputra A K, Prestwich G D, Cool S M, et al. Combining electrospun scaffolds with electrosprayed hydrogels leads to three-dimensional cellularization of hybrid constructs. Biomacromolecules, 2008, 9 (8): 2097-2103.

[118] Thorvaldsson A, Silva-Correia J, Oliveira J M, et al. Development of nanofiber-reinforced hydrogel scaffolds for nucleus pulposus regeneration by a combination of electrospinning and spraying technique. Journal of applied polymer science, 2013, 128 (2): 1158-1163.

[119] Ramakrishna S. An introduction to electrospinning and nanofibers. World scientific, 2005.

[120] Yoo H S, Kim T G, Park T G. Surface-functionalized electrospun nanofibers for tissue engineering and drug delivery. Advanced drug delivery reviews, 2009, 61 (12): 1033-1042.

[121] Boudriot U, Dersch R, Greiner A, et al. Electrospinning approaches toward scaffold engineering—a brief overview. Artificial organs, 2006, 30 (10): 785-792.

[122] Vatankhah-Varnosfaderani M, Daniel W F, Everhart M H, et al. Mimicking biological stress–strain behaviour with synthetic elastomers. Nature, 2017, 549 (7673): 497-501.

[123] Daniel W F, Burdyńska J, Vatankhah-Varnoosfaderani M, et al. Solvent-free, supersoft and superelastic bottlebrush melts and networks. Nature materials, 2016, 15 (2): 183-189.

[124] Vohidov F, Milling L E, Chen Q, et al. ABC triblock bottlebrush copolymer-based injectable hydrogels: design, synthesis, and application to expanding the therapeutic index of cancer immunochemotherapy. Chemical Science, 2020, 11 (23): 5974-5986.

[125] Erman B, Mark J E. Structures and properties of rubberlike networks. Oxford University Press, 1997.

[126] Mark J E, Erman B. Rubberlike elasticity: a molecular primer. Cambridge University Press, 2007.

[127] Zhou Y, Zhang W, Hu J, et al. The stiffness-threshold conflict in polymer networks and a resolution. Journal of Applied Mechanics, 2020, 87 (3): 031002.

第6章
水凝胶极限力学性能的设计原理和调控方法

6.1 韧性：在可拉伸高分子网络中引入能量耗散机制

6.2 强度：让高分子网络内部有足够多的分子链能够同时硬化且断裂

6.3 弹性：降低水凝胶在一定变形范围内的机械耗散

6.4 韧性黏结：整合具有机械耗散的增韧水凝胶基体与高强界面的交联

6.5 抗疲劳：用具有高本征断裂能的物质去阻碍疲劳裂纹扩展

6.6 抗疲劳粘接：在界面处强力固定具有高本征断裂能的物质

参考文献

在过去几十年中，开发了许多 UPN 架构和 UPN 相互作用，但具有极端力学性能的水凝胶的设计在很大程度上遵循了爱迪生的方法——对特定高分子进行反复试验。使用不同的候选材料和制造方法合理设计用于各种应用的水凝胶仍然是目前软材料领域的核心需求。在本章中，我们将给出水凝胶的通用设计原则，以实现相应的极端力学性能的构建，包括极高的断裂韧性、拉伸强度、回弹性、界面韧性、疲劳阈值和界面疲劳阈值。然后，我们将使用 UPN 架构和 / 或 UPN 相互作用讨论这些设计原则的实施策略。

6.1 韧性：在可拉伸高分子网络中引入能量耗散机制

6.1.1 断裂韧性

断裂韧性已被广泛用于表征材料在机械载荷下抵抗断裂的能力。材料断裂韧性通常用在未变形状态下测量的单位面积上传播材料裂纹所需的能量进行定量评估（图 6.1），可通过式（6.1）计算：

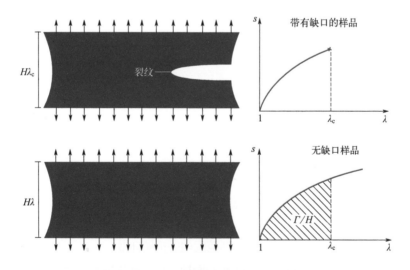

图 6.1　断裂韧性的定义以及测量断裂韧性的纯剪切试验示意图 [4]

$$\varGamma = G_c = -\frac{\mathrm{d}U}{\mathrm{d}A} \qquad (6.1)$$

式中，$\varGamma$ 是断裂韧性；U 是系统的总势能；A 是在未变形状态下测得的裂纹面积；G_c 是驱动裂纹扩展的临界能量释放率。根据式（6.1）所得的断裂韧性的单位是焦耳每平方米（J/m^2）。

弹性体和水凝胶等软材料的断裂韧性可以通过纯剪切试验和单缺口试验等方法进行测量[1-3]。在纯剪切试验中，两块具有相同的厚度 T、宽度 W 和高度 H 的水凝胶，且 $W \gg H \gg T$（图 6.1），分别用刚性板沿长边（即沿着宽度方向）夹住。在其中一个样品中引入一个长度约为 $0.5W$ 的凹口，逐渐将其拉伸至其未变形长度的 λ_c 倍，直到裂纹开始从凹口扩展（图 6.1）。另一个无凹口的样品被均匀拉伸到临界拉伸 λ_c 以上，以测量标称应力 s 与拉伸 λ 的关系（图 6.1）。此后，根据纯剪切试验中测得的 λ_c 和 s-λ 的关系，可以将水凝胶的断裂韧性表示为 $\Gamma = H\int_1^{\lambda_c} s d\lambda$。

正如第 5 章中所述，传统高分子网络的断裂韧性是其固有断裂能 Γ_0，这是在单位面积上断裂一层高分子链所需的能量（图 6.2）。依据传统高分子网络的典型参数进行评估，可以得出相应水凝胶的断裂韧性通常在每平方米几十焦耳以内。此外，具有传统高分子网络的水凝胶的断裂韧性还与它们的拉伸极限和剪切模量有关（参见第 5 章）。为了增加传统高分子网络的断裂韧性，需要增加链长（N），从而增加高分子网络的拉伸极限；然而，高分子网络的链密度（n）和剪切模量将降低。

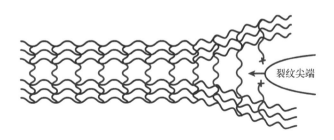

图 6.2　高分子链层断裂产生固有断裂能的示意图[4]

6.1.2　坚韧水凝胶的设计原则

坚韧水凝胶的设计原则（图 6.3）与各种工程材料（如金属[5]、陶瓷[6]、复合材料[7]、高分子[8]）和各种生物组织（如肌腱、软骨、肌肉、血管）[9] 的增韧机制一样，即将延展性和机械耗散整合在同一材料中，以便在裂纹扩展之前在裂纹尖端周围形成一个具有大量机械耗散的加工区 [图 6.3（a）]。材料的机械耗散表现为加载 - 卸载循环下其应力 - 应变曲线上的滞后回线 [图 6.3（a）]。水凝胶的延展性通常依赖于其高分子网络的高拉伸性（或高拉伸极限）。具体而言，坚韧水凝胶的设计原则是将耗散分散到弹性高分子网络中[1,10]。具有机械耗散能力的水凝胶的总断裂韧性可定量地表示为[1,11]：

$$\varGamma = \varGamma_0 + \varGamma_D \tag{6.2}$$

式中，$\varGamma$、$\varGamma_0$、$\varGamma_D$ 分别是总断裂韧性、固有断裂能、加工区机械耗散。虽然水凝胶的固有断裂能通常限制在每平方米几十焦耳，但加工区耗散的贡献可能非常高，因为加工区单位体积的耗散能量和加工区的尺寸都可以是很大的值 [图 6.3（c）]。事实上，目前构建出的坚韧水凝胶的断裂韧性已超过 $10000 J/m^2$，这比基于传统高分子网络的水凝胶高出好几个数量级 [1]。

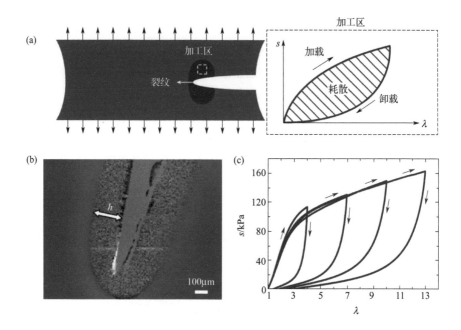

图 6.3 坚韧水凝胶的设计原则

（a）裂纹尖端周围加工区的机械耗散对总断裂韧性的贡献，机械耗散表现为应力－应变曲线上的滞后回线；（b）PAMPS-PAM 水凝胶裂纹尖端周围加工区的显微镜图像；（c）PAM-海藻酸盐水凝胶在加载和卸载循环下的标称应力 s 与拉伸 λ 的关系[4,12,13]

6.1.3 坚韧水凝胶的实施策略

在构建坚韧水凝胶时通常需满足以下两点：水凝胶中至少有一个高分子网络保持高拉伸极限，因此该高分子网络中的高分子链需要具有高 N 值；水凝胶中至少一种组分在加工区经历变形时耗散大量机械能。坚韧水凝胶的设计原则已通过各种类型的 UPN 架构和 UPN 相互作用实现。接下来，我们将讨论一些具体实例。

自 2003 年 Gong 等开创性地构建了双网络水凝胶之后，互穿高分子网络和半互穿高分子网络已被广泛用于坚韧水凝胶的设计（图 6.4）[14]。典型的双网络水凝胶会贯穿长链网络（高 N）和短链网络（低 N）[14]。当双网络水凝胶变形时，短链网络会断裂并消耗大量机械能，而长链网络即使在高拉伸下也能保持水凝胶的完整性，材料满足坚韧水凝胶的设计原则[10,13,14]。此外，他们不仅首次证明了双网络水凝胶的断裂韧性可以超过 $1000J/m^2$[14,15]；还为坚韧水凝胶[16]和弹性体[17]开发了三重网络结构等其他互穿和半互穿高分子网络。需要注意的是，由于短链网络的断裂通常是不可逆的，因此这些水凝胶的机械耗散能力在几次大变形循环后可能会显著降低[11]。

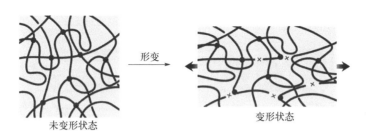

图 6.4　基于互穿或半互穿高分子网络的坚韧水凝胶的实施策略[4]

具有高官能交联的高分子网络提供了基于各种类型的高分子和高官能交联的坚韧水凝胶。两个相邻的高功能交联点之间有多个高分子链（超过 10 个）桥接，并且这些高分子链的长度通常是不均匀的（图 6.5）。当水凝胶变形时，相对较短的高分子链断裂或从高官能交联中分离，而相对较长的高分子链会保持水凝胶的完整性和高拉伸性，以满足坚韧水凝胶的设计原则。高分子链和高官能交联之间的键可以是永久共价交联[18]、强物理交联[19,20]、弱物理交联[21,22]、动态共价交联[23,24]或几者间的组合[25]。根据相邻交点之间高分子链的数量和长度以及高分子链和交联点之间的键类型，相应的水凝胶可以具有不同的机械耗散能力和拉伸能力，因此表现出不同的断裂韧性。

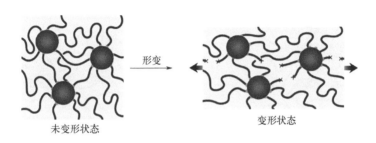

图 6.5　基于高官能交联高分子网络的坚韧水凝胶的实施策略[4]

由于自有的可拉伸性且在形变后的重新定向和重新排列的特点，微纳纤维所构筑的水凝胶材料具有较强的可拉伸性（图 6.6）[20,23,26-29]。此外，微纳纤维的断裂及其从水凝胶基质中的解离会消耗大量的机械能。兼具高拉伸性和机械耗散能力，使得微纳纤维高分子网络可用于构建坚韧水凝胶。

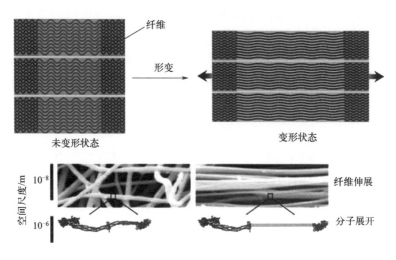

图 6.6　基于微纳纤维高分子网络的坚韧水凝胶的实施策略 [4,29]

除了上述 UNP 架构外，UPN 相互作用也应用于坚韧水凝胶的设计过程中 [30]。例如，晶畴和玻璃状结节等强物理交联可以轻易地实现相应 UPN 架构的高官能交联（图 6.5），并形成坚韧水凝胶。

弱物理交联 [31-40] 和动态共价交联 [23] 也应用于具有长高分子链的高分子网络中（即通过永久共价键稀疏交联的高分子网络）以设计坚韧水凝胶。弱物理交联和动态共价交联在这些水凝胶中充当可逆交联（图 6.7）。随着水凝胶的变形，这些可逆交联键会以解离或去交联形式耗散大量机械能，而稀疏交联的长链高分子网络能维持高分子网络的高拉伸性（图 6.7）。结合了可逆和共价交联实现的机械耗散和高拉伸性的协同作用实现了坚韧水凝胶的设计原理。

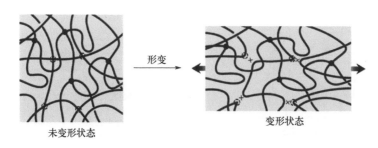

图 6.7　基于可逆交联高分子网络的坚韧水凝胶的实施策略 [4]

弱物理交联和动态共价交联也可应用于 UPN 架构，如互穿高分子网络（图 6.4）[12,41-49]、具有高官能交联的高分子网络（图 6.5）[19,21,22,50-53] 以及微纳纤维高分子网络（图 6.6）[26,27,51,54]，利用这些可逆键耗散额外机械能的能力，可以进一步强化所得的水凝胶。此外，与不可逆断裂的高分子链不同，解离的弱物理交联和动态共价交联会由于其可逆性而重新结合，从而赋予坚韧水凝胶在循环载荷下可恢复的耗散 [1,12]。

6.2 强度：让高分子网络内部有足够多的分子链能够同时硬化且断裂

6.2.1 抗拉强度

通常采用抗拉强度、抗压强度和剪切强度等多种强度测量方式来表征材料的强度。我们关注水凝胶的拉伸强度的原因有以下两点：①样品的拉伸、压缩和剪切变形相互关联。例如，样品的单轴压缩相当于样品的双轴拉伸；样品的纯剪切相当于样品在一个方向上被拉长并在垂直方向上被缩短。②抗拉强度比抗剪强度更容易测量，抗拉强度比抗压强度在测量中受边界条件（如摩擦）的影响更小。

由于弹性体和水凝胶等软材料通常不会塑性屈服，因此它们的拉伸强度通常定义为在单轴拉伸试验中发生最终拉伸失效时的应力。此外，由于水凝胶样品在失效前通常会发生大变形，因此可以根据标称应力或真实应力来定义拉伸强度（图 6.8），如下所示：

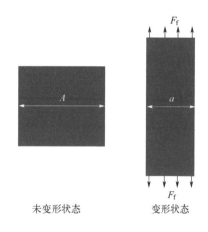

图 6.8 抗拉强度的定义和测量

$$s_\mathrm{f} = \frac{F_\mathrm{f}}{A}, \sigma_\mathrm{f} = \frac{F_\mathrm{f}}{a} \tag{6.3}$$

式中，F_f 是样品失效时的拉力；A、a 分别是参考（未变形）和拉伸（变形）状态下样品的横截面积；s_f、σ_f 分别是标称拉伸应力、真实拉伸应力。具有传统高分子网络的水凝胶即便是坚韧水凝胶的标称拉伸强度通常低于1MPa[12,14]，这远远低于金属等工程材料和肌腱等生物组织的拉伸强度。

6.2.2 抗拉伸水凝胶的设计原则

通常情况下，抗拉伸水凝胶设计原则是可以使高分子网络中的大量高分子链变硬的同时发生断裂（图6.9）。遵循这一原则，高分子网络的标称和真实拉伸强度分别可以评估为：

$$s_f = M_f f_f, \sigma_f = m_f f_f \tag{6.4}$$

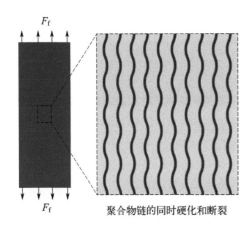

图6.9 多个高分子链的同时硬化和断裂的示意图[4,56]

式中，f_f是断裂单个高分子链所需的力，大约为几纳牛顿[55]；M_f、m_f分别是未变形和变形时单位面积高分子网络同时断裂的高分子链数。在高分子网络中所有高分子链同时断裂的理想情况下，s_f和σ_f可以分别达到1GPa和10GPa[56]。

在现实情况下，几乎所有材料都存在诸如缺口、微裂纹、空腔、杂质以及缺失高分子链或缺失交联等形式的缺陷，而缺陷的存在通常会显著降低材料的抗拉强度[57-59]。在不失一般性的情况下，假设拉伸样品中最大的缺陷是在未变形状态下垂直于拉伸方向的长度为D的缺口（图6.10）。样品的抗拉强度通常随着缺陷尺寸D的减小而增加，直至达到临界值D_c，低于该值时抗拉强度对缺陷不敏感（图6.10）。临界缺陷尺寸D_c可表示为[58,59]：

$$D_c \approx \frac{\Gamma}{W_c} \tag{6.5}$$

式中，W_c是缺陷不敏感样品单位体积的拉伸破坏功；Γ是样品的断裂韧性。

为了获得坚韧水凝胶，我们希望水凝胶样品具有对缺陷不敏感的拉伸强度[56]。根据式（6.5）可知，当临界缺陷尺寸越大时（即较大的D_c），越坚韧的材料（即具有较高Γ的材料）可能对较大的缺陷不敏感。例如，玻璃和陶瓷的临界缺陷尺寸为几纳米，脆性水凝胶的临界缺陷尺寸为几微米，坚韧弹性体和水凝胶的临界缺陷尺寸为几毫米[58,59]。此外，将样品的特征尺寸（如图 6.10 中的样品直径）设置为类似于或小于临界缺陷尺寸 D_c 是一种常见的策略，从而保证样品的抗拉强度对样品中任何可能存在的缺陷不敏感[56]。

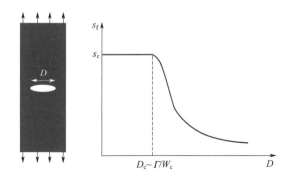

图 6.10 标称抗拉强度 s_c 会随着缺陷尺寸 D 的减小而增加，直至达到临界值 D_c，低于该值时抗拉强度对缺陷不敏感[4,58,59]

6.2.3 抗拉伸水凝胶的实施策略

具有高官能交联（如纳米晶域）的 UPN 已广泛用于抗拉伸水凝胶的设计[60]。水凝胶发生大变形时，相对较短的高分子链会逐渐从纳米晶域中被拉出，因此桥接相邻纳米晶域的高分子链往往具有相似的长度，进而变硬并断裂——实现强抗拉伸水凝胶设计原则（图 6.11）。

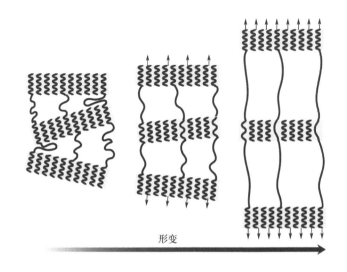

图 6.11 基于高官能交联高分子网络的抗拉伸水凝胶的实施策略[4]

微纳纤维高分子网络是另一种 UPN 架构。微纳纤维的直径可以很容易地控制在临界缺陷尺寸 D_c 以下，其中的高分子链束可以设计为变硬并断裂，以赋予纤维高拉伸强度或理想强度（图 6.12）[20,23,28,61,62]。因此，由此产生的微纳纤维水凝胶可以达到极高的拉伸强度（图 6.13）。根据第 1 章的介绍可知，肌腱、韧带和肌肉等生物水凝胶通常采用具有分层结构的纳米纤维和微米纤维，以实现高拉伸强度。

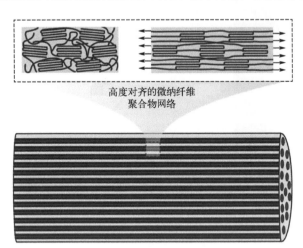

图 6.12　基于微纳纤维高分子网络的抗拉伸水凝胶的实施策略 [4]

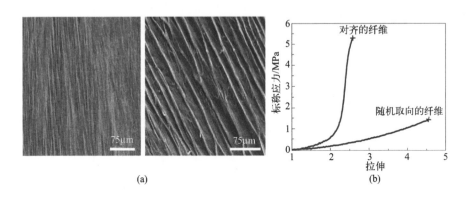

图 6.13　（a）具有对齐纤维的纤维 PVA 水凝胶的共聚焦（左）和 SEM（右）图像；（b）具有排列和随机取向纤维的纤维状 PVA 水凝胶的标称应力 - 拉伸曲线 [61]

除了上述 UPN 架构外，UPN 相互作用也可以促进强抗拉伸水凝胶设计原则的实施。诸如晶畴之类的强物理交联允许从中拉出高分子链，以实现多个高分子链的同时硬化和断裂（图 6.9）[60]。而氢键等弱物理交联可以促进高分子链的排列和自组装成束，这往往能使得这些高分子链变硬并断裂，从而使水凝胶具有高拉伸强度。

在结构层面上，由高分子[63,64]、钢[65]、玻璃[66,67]、木材[68]等制成的高强度粗纤维已被用于增强水凝胶，所得水凝胶的拉伸强度主要由粗纤维的强度决定。

6.3 弹性：降低水凝胶在一定变形范围内的机械耗散

6.3.1 弹性

弹性体和水凝胶等软材料的弹性通常定义为变形恢复中释放的能量与引起材料变形所需的能量之比[69]。以加载-卸载循环中进行单轴拉伸试验的圆柱形样品为例（图6.14），若卸载中释放的能量和样品单位体积耗散的能量分别表示为W_R和W_D。那么材料的回弹性R和滞后比H可分别表示为[69,70]：

$$R = \frac{W_R}{W_R + W_D}, H = \frac{W_D}{W_R + W_D} = 1 - R \tag{6.6}$$

式中，回弹性R和滞后比H取决于材料的特性和加载的条件（如施加的拉伸力和拉伸率）。通常，软材料的回弹性可以通过循环拉伸试验和落球试验等多种实验方法进行测量[69,70]。

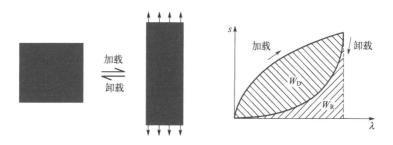

图6.14 弹性的定义和测量示意图[4]

6.3.2 高弹性水凝胶的设计原则

一旦材料变形断裂，储存在材料中的大部分弹性能量（因形变而引起的势能）会被耗散[70-72]，能量恢复率低，因此断裂材料的回弹力低。因此，水凝胶的高回弹性只能设计到水凝胶的断裂情形。弹性水凝胶设计的一般原则是在水凝胶经历的一定形变的范围内尽量减少水凝胶的机械耗散，或者简而言之，延迟耗散[70]。在不失一般性的情况下，我们定义了水凝胶中高分子链的临界拉伸（λ_R），低于该临界拉伸时，水凝胶可以在形变恢复期间释放大部分储存的弹性能量（即 $W_D \approx 0$，图 6.15）[70]。因此，根据式（6.6），水凝胶在满足式（6.7）的条件下将具有高弹性：

$$\lambda \leqslant \lambda_R \leqslant \lambda_{\lim} \tag{6.7}$$

式中，λ 和 $\lambda_{\lim}$ 分别是水凝胶中高分子链的拉伸和拉伸极限。

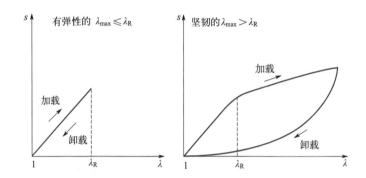

图 6.15　当拉伸低于临界拉伸 λ_R 时，水凝胶在变形恢复过程中释放大部分储存的弹性能量，具有高弹性；当拉伸高于 λ_R 时，水凝胶会消耗大量机械能，从而提供高断裂韧性[4]

弹性水凝胶的设计原则还调和了一对看似矛盾的特性，即断裂韧性和弹性。水凝胶在 $\lambda \leqslant \lambda_R$ 的适度变形下具有高弹性（图 6.15）；然而，当裂纹试图在水凝胶中扩展时，裂纹尖端周围加工区的链拉伸可能远高于 λ_R，从而引起大量机械耗散以增韧水凝胶（图 6.16）。事实上，心脏瓣膜等生物水凝胶能够将机械耗散延迟到超生理变形水平，以实现高断裂韧性和高弹性[73,74]。合成弹性体[75]、水凝胶[70] 以及水凝胶复合材料[76] 均可通过延迟耗散的设计原则变得兼具坚韧和弹性（图 6.17）。

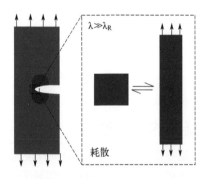

图 6.16　弹性水凝胶裂纹周围加工区的拉伸通常远高于 λ_R，耗散大量机械能并提供高断裂韧性[4]

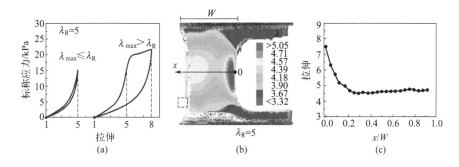

图 6.17 （a）$\lambda_R=5$ 的 PAM- 海藻酸盐水凝胶的标称应力与拉伸曲线；（b）$\lambda_R=5$ 的 PAM- 海藻酸盐水凝胶中裂纹周围的测量变形；（c）加工区的拉伸可以远高于 $\lambda_R=5$ 条件下的拉伸[70]

6.3.3 高弹性水凝胶的实施策略

理想高分子网络是一种用于实现弹性水凝胶的设计原理的常见 UPN 架构[71,72]。由于理想高分子网络中的高分子链具有相对均匀的长度且没有缠结，因此具有理想高分子网络的水凝胶在达到拉伸极限前可以变形且通常不会产生明显的机械耗散，从而产生高弹性（图 6.18）[71,72]。此外，具有可滑动交联点的高分子网络也能够实施弹性水凝胶的设计原理，因为在其重构过程中滑动交联所消耗的能量低得可以忽略不计。尽管理想高分子网络和具有可滑动交联点的高分子网络具有弹性，但它们并不坚韧，因为它们的断裂韧性仍然是用于使高分子链层断裂的固有断裂能 Γ_0。

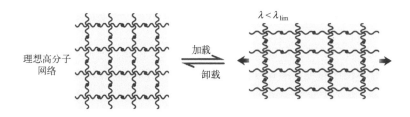

图 6.18 理想的高分子网络在断裂之前具有弹性[4]

包括互穿高分子网络、半互穿高分子网络和具有高官能交联的高分子网络在内的多峰高分子网络。由于高分子网络中的短高分子链的断裂和/或去交联，通常在非常小的变形时就开始耗散机械能，这种"早期"耗散使得水凝胶在实际应用中仅可实现窄范围的弹性变形[70]。为了解决"早期"耗散的问题，Lin 等将互穿高分子网络预拉伸至 λ_R 以断裂和/或去交联易受影响的短高分子链，从而减少 λ_R 变形范围内可能的耗散机制（图 6.19）[70]。在随后的测试中，如果水凝胶中

的链拉伸低于 λ_R，则水凝胶具有高弹性，因为在此范围内水凝胶缺乏机械耗散机制。当高分子链拉伸超过 λ_R 时，在裂纹尖端周围加工区的一些高分子链将进一步断裂和去交联以耗散机械能并使水凝胶变韧（图 6.16、图 6.17 和图 6.19）。此外，其他的多峰高分子网络，如半互穿高分子网络和具有高官能交联的高分子网络，可以以类似的方式预拉伸，从而实现弹性水凝胶的设计原理（图 6.20）[77]。需要注意的是，当预拉伸多峰高分子网络时，高分子链的断裂和去交联应该是不可逆的，因此耗散机制一旦耗尽就无法恢复[61]。

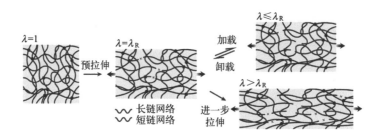

图 6.19　预拉伸互穿高分子网络到 λ_R，可以使它们兼具弹性和韧性[4]

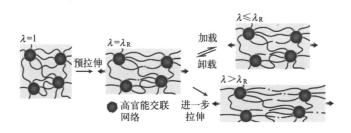

图 6.20　预拉伸高官能交联高分子网络到 λ_R，可以使它们兼具弹性和韧性[4]

微纳纤维高分子网络可以通过弹性微纳纤维来构建相应的水凝胶，从而实现弹性水凝胶的设计原理（图 6.21）[61]。此外，由于断裂和拉出微纳纤维所需的能量可能远高于断裂无定形高分子链所需的能量，因此弹性微纳纤维水凝胶也可能很坚韧[61]。

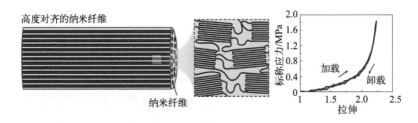

图 6.21　具有弹性纤维的微纳纤维高分子网络兼具弹性和韧性[4,61]

除了上述 UPN 架构外，一些 UPN 相互作用也可以促进弹性水凝胶设计原则的实施。诸如晶畴之类的强物理交联可以为某些 UPN 架构提供高官能交联，这同样可以通过预拉伸来构建弹性水凝胶[60]。需要注意的是，弱物理交联和动态共价交联因具有可逆性和耗散性，可能不适合实施弹性水凝胶的设计原理[70]。

在结构层面上，可以将弹性粗纤维嵌入弹性水凝胶基质中，以构架兼具弹性和韧性的水凝胶复合材料[76]。

6.4 韧性黏结：整合具有机械耗散的增韧水凝胶基体与高强界面的交联

6.4.1 界面韧性

界面韧性，又称为实际黏附功，通常用于表征两种黏附材料的界面在机械载荷下抵抗断裂的能力。两种黏附材料之间的界面韧性的一个常见定义是：在材料未变形状态下测量的单位面积上沿界面或任一材料传播裂纹所需的能量 [图 6.22（a）][78]。根据裂纹是沿界面还是沿任一材料传播，将相应的失效模式分为黏合失效和内聚失效。界面韧性 Γ^{inter} 可以定量表示为：

$$\Gamma^{\text{inter}} = G_c = -\frac{dU}{dA} \tag{6.8}$$

式中，U 是系统的总势能；A 是在未变形状态下测得的裂纹面积；G_c 是驱动界面裂纹扩展的临界能量释放率。根据式（6.8）可知，界面韧性的单位是焦耳每平方米（J/m^2）。

弹性体和水凝胶等软材料的界面韧性可以通过 90° 剥离试验、T 形剥离试验和搭接剪切试验等多种方法进行测量[2,78]。例如，在 90° 剥离试验中，将一层厚度为 T，宽度为 W，长度为 $L(L \gg W \gg T)$ 的水凝胶黏合在基材上，并在界面上引入沿长度方向的缺口 [图 6.22（a）]；随着施加外力的增大，水凝胶的分离部分进一步从基底上剥离，同时保持垂直于基底。当剥离过程进入稳态时，测得的力达到平台力 F_{plateau}，界面韧性通常通过将平台力 F_{plateau} 除以水凝胶片的宽度 W 来确定，即 $\Gamma^{\text{inter}} = F_{\text{plateau}}/W$。

如果具有传统高分子网络的水凝胶以共价键等方式牢固地结合在基材上，则界面韧性处于水凝胶的断裂韧性或固有断裂能 Γ_0 的水平。这是因为水凝胶的断裂韧性构成了界面韧性的上限,当超过该上限时可能会发生内聚破坏模式 [图 6.22（c）][79]。例如,用传统高分子网络的典型参数进行评估时,水凝胶的界面韧性在每平方米几十焦耳以内。如果水凝胶通过低密度的弱物理交联（如氢键、静电相互作用等）黏附在基材上,则界面韧性可能会更低,因为该情况下可能会发生黏附失效模式 [图 6.22（b）][80]。

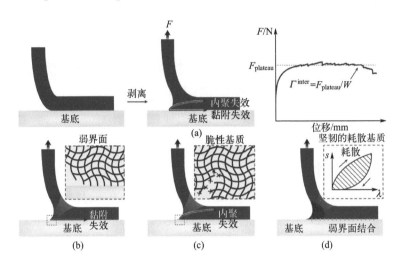

图 6.22　水凝胶的界面韧性

（a）界面韧性的定义以及测量界面韧性的 90° 剥离试验。F 是剥离力,$F_{plateau}$ 是平台力,W 是样品的宽度。在 90° 剥离试验中测量的 $F_{plateau}$ 和 W 的值,界面韧性可以计算为 $\Gamma^{inter}=F_{plateau}/W$。（b）弱黏合界面的黏合剂失效模式。（c）脆性水凝胶基质的内聚黏附失效模式。（d）坚韧耗散水凝胶和强界面连接的结合使水凝胶具有坚韧的黏附力[4,79]

6.4.2　强界面黏附性水凝胶的设计原则

如上所述,如果水凝胶通过低密度的弱物理交联键黏附到基材上,裂纹很容易沿着水凝胶 - 基材界面传播,导致界面韧性低 [图 6.22（b）]。因此,水凝胶坚韧黏合的设计首先需要水凝胶和黏合基材之间的强界面连接,例如共价键 [79,81,82]、强物理交联 [80,83,84]、连接高分子 [77,85-87] 和机械互锁等 [88,89]。此外,由于界面裂纹可以延伸到块状水凝胶中并引起内聚式破坏 [图 6.22（c）],因此水凝胶的强黏附设计需要进一步提升水凝胶基质的断裂韧性 [79]。

具体而言,水凝胶坚韧黏附的设计原则是将坚韧的耗散水凝胶基质与强界面键合相结合 [79]。当外力试图将坚韧水凝胶从基材上分离出来时,强大的界面结合将保持界面裂纹尖端,而使整个水凝胶形成一个具有大量机械耗散的加工区 [图 6.22（d）]。总界面韧性可定量表示为 [79,82]：

$$\varGamma^{\text{inter}} = \varGamma_0^{\text{inter}} + \varGamma_D^{\text{inter}} \tag{6.9}$$

式中，$\varGamma^{\text{inter}}$、$\varGamma_0^{\text{inter}}$、$\varGamma_D^{\text{inter}}$ 分别是总界面韧性、由强界面连接而产生的固有界面韧性、加工区机械耗散所引起的界面韧性。

动物体内的生物水凝胶（如软骨、肌腱和骨骼上的韧带）的黏附性，同样依赖于坚韧水凝胶和强界面键合的结合。然而，由于在之前的研究中坚韧的耗散水凝胶基质在水凝胶黏附方面的作用尚未得到充分探索或被低估[79,92,93]，直到最近才实现了合成水凝胶在不同基材（包括金属、陶瓷、玻璃、硅树脂、弹性体、水凝胶和生物组织）上的强黏附性[77,79,81,85-87,90,91]。值得注意的是，强界面键合和/或黏附体的大量耗散也被广泛用于工程材料（如金属[94]和橡胶[95,96]）在基材上牢固黏合的设计中。

6.4.3 强界面黏附性水凝胶的实施策略

由于在 6.1 节中已经讨论了坚韧水凝胶的实现手段，本节中我们将重点关注如何实现强界面键合从而在各种基材上黏合坚韧的耗散水凝胶。为了实现坚韧的附着力，界面连接的固有界面韧性 $\varGamma_0^{\text{inter}}$ 至少应达到坚韧水凝胶的固有断裂能 $\varGamma_0$ 的水平，即每平方米几十焦耳以上[79]。鉴于对固有界面韧性的这一要求，强界面连接通常通过共价键、强物理交联、连接高分子和机械互锁来实现（图 6.23）。

共价键粘接是最广泛使用的手段之一，以将坚韧水凝胶 UPN 中的高分子链牢固地固定在各种基材上。用于水凝胶强黏附的常用共价键包括碳-碳、碳-氮、碳-硫、碳-氧和硅-氧键[97]。为了形成这些共价键，水凝胶和基底表面通常修饰上对应的官能团，如可交联的碳碳不饱和键（形成碳-碳键）[98]，氨基（形成碳-氮键）[79]，硫醇基（形成碳-硫键）[99]，羟基和羧基（形成碳-氧键）以及硅醇基（形成硅-氧键）[图 6.23（a）][100]。根据 Lake-Thomas 模型，共价锚定在基材上的高分子链的固有界面韧性 $\varGamma_0^{\text{inter}}$ 可表示为：

$$\varGamma_0^{\text{inter}} = M^{\text{inter}} N U_f \tag{6.10}$$

式中，M^{inter} 是在未变形参考状态下单位面积上共价锚定的高分子链的数量；N 是每个高分子链的库恩单体数量；U_f 是断裂库恩单体所需能量或底物上的共价键两者中的较低值。根据式（6.10），在基材上锚定具有更高共价键密度的长高分子链将提供更高的固有界面韧性值[79,82]。

包括晶畴、玻璃状结节和高密度物理键（如氢键）在内的强物理交联也可以将坚韧水凝胶牢固地黏附在基材上 [图 6.23（b）][37,83,99,100,103-109]。由于晶畴和玻璃状结节通常充当高官能交联，它们都可以将多个高分子链锚定在基板上，从而进一步增强固有界面韧性 Γ_0^{inter}。

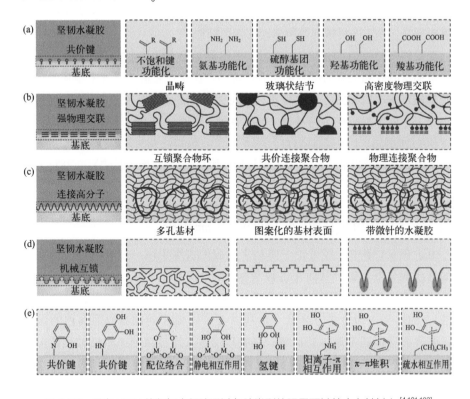

图 6.23　具有 UPN 的坚韧水凝胶通过各种类型的强界面键结合在基材上[4,101,102]
（a）共价键；（b）强物理交联；（c）连接高分子；（d）机械互锁；（e）儿茶酚相互作用可以实现各种类型的强界面连接

连接高分子同样被用于实现基材与黏合弹性体/水凝胶的牢固粘接 [图 6.23（c）][95,110]。在这种情况下，被粘接的基材通常以高分子网络（即弹性体和水凝胶）的形式存在。为了提供强界面连接，连接高分子可以与水凝胶和基底的高分子网络形成共价交联[77,85]、互锁环[86,87,99,105,111]或强物理交联[87]，由两个高分子网络中的单体聚合而成[85,86,100]，或直接将外来分子链添加到两个高分子网络的界面上[87]。

坚韧水凝胶和基底之间的机械互锁通常发生在从微米到毫米的尺度上[图 6.23（d）]。一种常用的方法是将坚韧水凝胶的前体溶液倒入多孔基材中，然后形成与基材机械互锁的坚韧水凝胶[88]。此外，还可以对基材表面进行粗糙化或图案化，以增强其与坚韧水凝胶的机械互锁强度[89,112,113]。作为一个特殊而有趣的案例，水凝胶可以被制成干燥的微针，从而能够刺入柔软的基质中（如生物组织），然后吸水膨胀形成机械互锁[114]。

受贻贝黏附蛋白的启发，儿茶酚化学已被广泛采用以实现水凝胶和基底之间的各种类型的界面连接 [图 6.23（e）][84,115]。儿茶酚可以与各种官能团形成共价或物理交联，氧化成醌后，儿茶酚可通过迈克尔加成与亲核试剂（例如胺和硫醇）形成共价键，并与金属氧化物形成强配位络合物[102]。儿茶酚的羟基不仅可以与金属氧化物形成静电相互作用，还可以与亲水性底物形成氢键；儿茶酚的苯环可与带正电荷的官能团形成阳离子-π相互作用，与苯环形成π-π堆积，与底物上的疏水官能团形成疏水相互作用[84,102]。虽然儿茶酚化学已被广泛用于水凝胶与基材之间的黏附，但仅通过基于儿茶酚的界面连接实现的黏附界面的韧性并不高[116]，这强调了坚韧耗散水凝胶基质对实现坚韧界面黏附的重要性[117]。

6.5 抗疲劳：用具有高本征断裂能的物质去阻碍疲劳裂纹扩展

6.5.1 疲劳阈值

"疲劳"一词被用于描述材料在长时间负载下观察到的许多症状，包括在长时间静态或循环负载下有或没有预切裂纹的材料[118,119]。在本节中，我们将重点关注具有预切裂纹的水凝胶在循环载荷下的疲劳断裂（图 6.24），因为这是水凝胶在机械动态环境中，如人造软骨[120]和软机器人[85]，最常见的失效模式之一。疲劳阈值 Γ_{FT}，定义为在无限载荷循环下发生疲劳裂纹扩展的最小断裂能[121,122]，用于评估描述材料在循环载荷下对疲劳裂纹扩展的抵抗，可以定量地表示为：

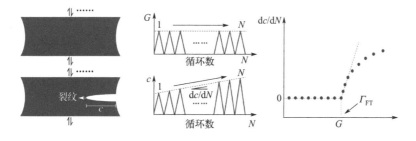

图 6.24 疲劳阈值的定义，以及用纯剪切法测量水凝胶的疲劳阈值[4]

$$\varGamma_{FT} = G_c(\mathrm{d}c/\mathrm{d}N \to 0) \tag{6.11}$$

式中，G 是在每个载荷循环下驱动裂纹扩展的能量释放速率；G_c 是在无限载荷循环（即 $\mathrm{d}c/\mathrm{d}N \to 0$）下裂纹扩展发生的最小能量释放速率；$c$ 是未变形状态下的裂纹长度；N 是施加载荷的循环次数。$\mathrm{d}c/\mathrm{d}N$ 可以给出每个循环的裂纹扩展。

弹性体和水凝胶等软材料的疲劳阈值可以通过纯剪切疲劳断裂试验和单缺口疲劳断裂试验等方法测量[60]。在纯剪切疲劳断裂试验中，两块具有相同厚度 T、宽度 W、高度 H（其中 $W \gg H \gg T$）的水凝胶（图 6.24），沿着长边夹住。对于第一块样品，反复拉伸至其未变形高度的 $\lambda_{applied}$ 倍的拉伸，以测量标称应力 s 与拉伸 λ 的关系，相应的能量释放率可以计算为 $G = H \int_1^{\lambda_{applied}} s \mathrm{d}\lambda$，它是循环数 N 的函数（图 6.24）。对于第二块样品，引入长度约为 $0.5W$ 的缺口后，反复拉伸至相同的拉伸度 $\lambda_{applied}$，以测量裂纹长度 c 作为循环数 N 的函数。纯剪切疲劳断裂的循环测试是通过改变对施加的拉伸 $\lambda_{applied}$ 的值（即不同的能量释放率 G）实现的，该测试可以获得 $\mathrm{d}c/\mathrm{d}N$ 与 G 的曲线（图 6.24），通过将 $\mathrm{d}c/\mathrm{d}N$ 与 G 的曲线与 G 轴相交（即当 $\mathrm{d}c/\mathrm{d}N \to 0$ 时）来确定疲劳阈值 $\varGamma_{FT}$。需要注意的是，水凝胶的疲劳断裂试验通常在水性环境中进行，以避免水凝胶出现因长时间负载而脱水的情况[60,61]。

6.5.2 抗疲劳水凝胶的设计原则

正如 6.1 节中所述，可以通过将机械耗散构建到弹性高分子网络中来设计坚韧水凝胶[1]，因为裂纹尖端周围加工区的机械耗散可以显著提高水凝胶的断裂韧性。然而，不可逆耗散机制（例如加工区中高分子链的断裂）通常在循环载荷下会耗尽。即使是可逆交联等可逆耗散机制，一旦被耗尽通常也难以及时恢复以抵抗未来载荷循环中的疲劳裂纹扩展（图 6.25）[121-123]。因此，水凝胶和弹性体的疲劳阈值是它们的固有断裂能：

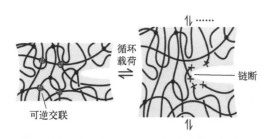

图 6.25 机械耗散机制，如坚韧水凝胶中的可逆交联，在循环载荷下会耗尽，不会影响疲劳阈值[4,60]

$$\varGamma_{FT} = \varGamma_0 \qquad (6.12)$$

根据式（6.12）可知，抗疲劳水凝胶的设计通常不能依赖于大块水凝胶基质中的机械耗散。

抗疲劳水凝胶的设计原理是使疲劳裂纹相遇到单位面积能量远高于单层高分子链的断裂能量的断裂物体，或者简单地说，通过固有的高能相固定疲劳裂纹（图 6.26）[60]。用于设计抗疲劳水凝胶的固有高能相包括纳米晶域 [图 6.27（a）][60]、微纳纤维 [图 6.27（b）][61]、微相分离 [图 6.27（c）][119,124]和宏观复合材料 [图 6.27（d）][76]。此外，由于抗疲劳水凝胶的设计不依赖于大块水凝胶的机械耗散，因此抗疲劳水凝胶通常表现出低滞后比 H 和高回弹性 R [式（6.6）][60,61,125]。值得注意的是，肌肉、肌腱和韧带等生物水凝胶通常具有分层排列的微纳纤维作为固有的高能相，以实现高疲劳阈值（参见第 1 章）。

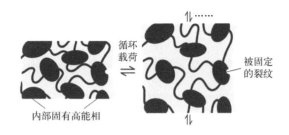

图 6.26　抗疲劳水凝胶通过其内部固有的高能相来固定疲劳裂纹 [4,60]

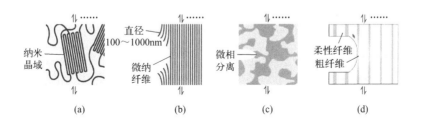

图 6.27　抗疲劳水凝胶的裂纹可以被固有高能相固定，包括（a）纳米晶域，（b）微纳纤维，（c）微相分离，（d）宏观复合材料 [4,60]

6.5.3　抗疲劳水凝胶的实施策略

可以通过具有固有高能相的 UPN 实施抗疲劳水凝胶的设计原则 [60,61,76,119,124]。为了有效地固定疲劳裂纹，UPN 中固有高能相的密度应该足够高 [60]。

晶畴等高官能交联可以有效地发挥 UPN 中固有高能相的作用 [图 6.27（a）]。

从晶畴中拉出高分子链所需的能量可能是断裂同一高分子链所需的能量的数倍，而机械破坏晶畴所需的能量可能是断裂相应的无定形高分子链所需的能量的数倍（图 6.28）[83]。研究表明，通过对水凝胶进行干法退火将 PVA 水凝胶的结晶度从 0.2%（质量分数）提高到 18.9% 可以将其疲劳阈值从 10J/m² 提高到 1000J/m²，达到软骨等抗疲劳生物水凝胶的水平[60]。由于 PVA 水凝胶中晶畴的尺寸仅为几纳米，因此晶畴起着固有高能相的作用 [图 6.27（a）]。此外，其他具有足够高密度的高官能交联的 UPN，如玻璃状结节，也可以实现抗疲劳水凝胶的设计原理。需要注意的是，具有高密度刚性晶畴和玻璃状结节的水凝胶可能会比普通水凝胶更硬[60]，然而这种高硬度对水凝胶的应用产生负面影响。

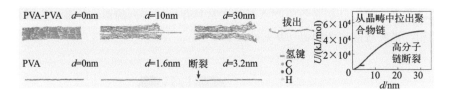

图 6.28　将高分子链拉出 PVA 晶畴并使同一高分子链断裂的分子动力学模拟[83]

微纳纤维也可以充当 UPN 中的固有高能相，以实施抗疲劳水凝胶的设计原理 [图 6.27（b）]。 由于纤维中成束高分子链的协同伸长和硬化，微纳纤维断裂所需的能量可能比相应的无定形高分子链断裂所需的能量高得多[61]。研究表明，通过冻融水凝胶将纳米纤维引入 PVA 水凝胶可以将其疲劳阈值从 10J/m² 提高到 310J/m²。如果通过预拉伸水凝胶使纳米纤维垂直于疲劳裂纹排列，则测得的疲劳阈值会进一步增加到 1250J/m²（图 6.29）[61]。此外，由于纳米纤维可以通过使用低密度的纳米晶域来缠结高分子链而使其变得柔顺、可拉伸和坚固，因此所得的纳米纤维水凝胶集高顺应性、可拉伸性和强度以及高疲劳阈值于一身[61]。

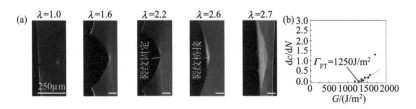

图 6.29　（a）微纳纤维 PVA 水凝胶中由纳米纤维固定的裂纹的共聚焦显微镜图像；（b）微纳纤维 PVA 水凝胶疲劳阈值的测量[61]

水凝胶中的相分离也可以提高水凝胶的疲劳阈值[119,124]，可能是因为破坏分离的相所需的能量高于破坏相应的无定形高分子链所需的能量。可逆共价键和弱物理交联等 UPN 相互作用在诱导水凝胶的相分离中可以起到关键的作用[119,124]。

在结构层面上，可以将宏观弹性体纤维嵌入弹性水凝胶中，形成宏观复合材料[76]。由于与一层无定形高分子链相比，它需要更高的能量来破坏弹性纤维，因此宏观复合材料的疲劳阈值超过 1000J/m²[图 6.27（d）][76]。

6.6 抗疲劳粘接：在界面处强力固定具有高本征断裂能的物质

6.6.1 界面疲劳阈值

黏附材料的界面可能会在长时间的载荷下发生疲劳失效，包括在长时间的静态或循环载荷下有或没有预切裂纹的界面。在本节中，我们将重点关注黏附在基底上的水凝胶在循环载荷下的疲劳断裂（图 6.30）。根据疲劳裂纹是沿界面传播还是在循环载荷下在水凝胶中传播，可以将对应的失效模式分为黏合失效或内聚失效[图 6.30（a）][83]。界面疲劳阈值通常用于表征黏附材料在循环载荷下任一失效模式下抵抗界面疲劳裂纹扩展的能力。界面疲劳阈值定义为在无限载荷循环下界面裂纹扩展发生的最小断裂能量[83,126-128]。与疲劳阈值类似，界面疲劳阈值 $\varGamma_{\mathrm{FT}}^{\mathrm{inter}}$ 可表示为：

$$\varGamma_{\mathrm{FT}}^{\mathrm{inter}} = G_{\mathrm{c}}\left(\mathrm{d}c/\mathrm{d}N \to 0\right) \tag{6.13}$$

式中，G 是在每个载荷循环下驱动界面裂纹扩展的能量释放速率；G_c 是在无限载荷循环（即 $\mathrm{d}c/\mathrm{d}N \to 0$）下界面裂纹扩展发生的最小能量释放速率；$c$ 是裂纹的长度；N 是施加载荷的循环次数。$\mathrm{d}c/\mathrm{d}N$ 可以提供每个循环的裂纹扩展情况。

弹性体和水凝胶等软材料的界面疲劳阈值可以通过循环 90°剥离试验、循环 T 形剥离试验和循环搭接剪切试验等方法进行测量[83,126-128]。在循环 90°剥离试验中[83,128]，一层厚度 T、宽度 W、长度 $L(L \gg W \gg T)$ 的水凝胶被黏合在基材上，并在界面上引入沿长度方向的缺口[图 6.30（a）]。在水凝胶的分离部分重复施加力 F，同时保持分离部分垂直于基底。施加的力 F 给出能量释放率 $G = F/W$，其中 W 是水凝胶片的宽度。测量界面裂纹长度 c，作为循环数 N 的函数。对于不同的施加力 F 值（即不同的能量释放率 G）重复 90°循环剥离试验，可以获得 $\mathrm{d}c/\mathrm{d}N$ 与 G 的关系曲线[图 6.30（a）]。通过将与 $\mathrm{d}c/\mathrm{d}N$ 的曲线在 G 轴的交点（当

$dc/dN \to 0$ 时)来确定界面疲劳阈值 Γ_{FT}^{inter}。需要注意的是,水凝胶的界面疲劳断裂试验也需要在水性环境中进行,以避免水凝胶因长时间负载而脱水 [83]。

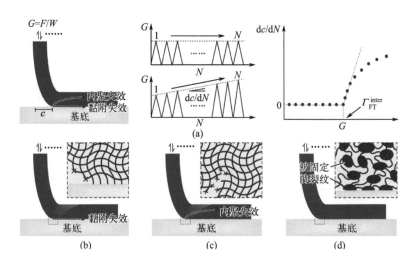

图 6.30 (a)界面疲劳阈值的定义,以及用于测量界面疲劳阈值的 90°循环剥离试验。F 是施加的剥离力,W 是样品的宽度,G 是能量释放率,c 是裂纹长度,N 是循环数。界面疲劳阈值 Γ_{FT}^{inter} 通过将 dc/dN 与 G 的曲线与 G 轴相交来确定。(b)沿界面的疲劳裂纹传播引起的黏合剂失效模式。(c)由于水凝胶中的疲劳裂纹扩展引起的内聚失效模式。(d)由界面上和本体水凝胶中固有高能相固定的疲劳裂纹 [4,83]

6.6.2 抗疲劳黏附水凝胶的设计原则

如第 6.4 节所述,水凝胶在基材上的坚韧黏附依赖于坚韧耗散水凝胶基质的构建以及强界面连接效应两者的结合 [79]。强大的界面键合可以保持界面裂纹尖端,而裂纹尖端周围加工区的机械耗散可以显著提高界面黏合的总界面韧性。然而,与水凝胶疲劳断裂的情况类似 [121-123],由于大块水凝胶基质中的机械耗散通常会因循环载荷而耗尽,因而这种耗散通常无法抵抗界面疲劳裂纹的扩展 [图 6.30(b)~(d)] [83,126]。因此,水凝胶和弹性体的界面疲劳阈值可以由它们的固有界面韧性来表示 [83,126]:

$$\Gamma_{FT}^{inter} = \Gamma_{0}^{inter} \qquad (6.14)$$

由于界面裂纹可以倾斜到块状水凝胶中并形成内聚破坏模式 [图 6.30(c)],水凝胶的抗疲劳黏附设计首先需要具有足够高密度的固有高能相的抗疲劳水凝胶基质 [60]。需要注意的是,仅坚韧但不耐疲劳的水凝胶基质不适合用于设计耐疲劳黏附，因为循环载荷会耗尽机械耗散。为了避免循环载荷下的黏合剂失效[图6.30（b）]，界面上的疲劳裂纹也需要由界面上牢固结合的固有高能相固定[图6.30（d）]。

简而言之，水凝胶抗疲劳黏附的设计原则是在界面上牢固地结合固有高能相[83]。虽然已有晶畴[83]和长高分子链[127]被用于设计抗疲劳黏合的固有高能相，但可以探索微纳纤维等其他候选材料。因为肌腱、韧带和软骨在内的生物水凝胶都是依赖于在它们与骨骼的界面上牢固结合的晶畴以及微纳纤维来实现抗疲劳黏附的[37]。

6.6.3 抗疲劳黏附水凝胶的实施策略

如6.5节中所述，抗疲劳水凝胶的设计依赖于在UPN中实现足够高的固有高能相密度。在本节中，我们将重点关注如何将固有高能相牢固地键合在基板上，以实现抗疲劳黏合的设计原理。

晶畴等高官能交联可以在UPN中发挥固有的高能相的作用。PVA水凝胶中的晶畴可以通过高密度氢键牢固地结合在各种基材上，包括玻璃、陶瓷、金属和弹性体（图6.31）[83]。分子动力学模拟表明，将高分子链拉出界面所需的能量远高于断裂相同的高分子链或将相同高分子链拉出晶畴所需的能量，这意味着 $\varGamma_0^{\text{inter}}$ 的高固有界面韧性[83]。因此，在磷酸盐缓冲盐水中测量的PVA-基底黏附系统的界面疲劳阈值高达 800J/m^2，类似于肌腱/韧带/软骨-骨界面的界面疲劳阈值（图6.32）。此外，在界面疲劳断裂试验中观察到的PVA-基底黏附系统的失效模式遵循内聚失效，表明了抗疲劳水凝胶中的固有高能相对于抗疲劳黏附设计的关键作用[83]。

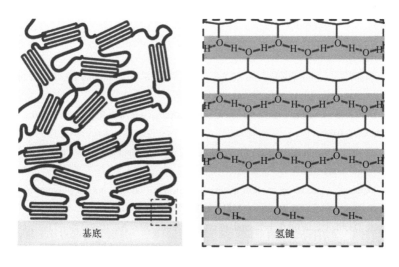

图6.31 通过高密度氢键实现基于UPN的抗疲劳的黏附水凝胶[4]

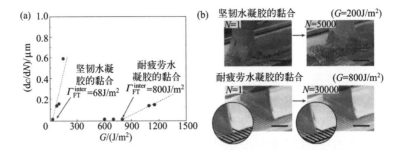

图6.32 （a）水凝胶在基材上的强黏附力和耐疲劳黏附力的疲劳阈值测量；（b）水凝胶在基材上的强黏附力（上）和耐疲劳黏附力（下）循环剥离试验中界面裂纹扩展的照片[83]

共价键也能够牢固地结合固有高能相（如基板上的纳米晶域和微纳纤维）（图6.33）。在多孔、粗糙或图案化的基材上固化抗疲劳水凝胶的前体溶液可实现机械互锁，从而在水凝胶-基材界面上强烈结合固有高能相（图6.33）[129]。

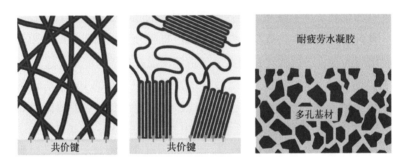

图6.33 通过高密度共价键或机械互锁实现基于UPN的抗疲劳的黏附水凝胶[4]

参考文献

[1] Zhao X. Multi-scale multi-mechanism design of tough hydrogels：building dissipation into stretchy networks. Soft matter，2014，10（5）：672-687.

[2] Creton C，Ciccotti M. Fracture and adhesion of soft materials：a review. Reports on Progress in Physics，2016，79（4）：046601.

[3] Long R，Hui C Y. Fracture toughness of hydrogels：measurement and interpretation. Soft Matter，2016，12（39）：8069-8086.

[4] Zhao X，Chen X，Yuk H，et al. Soft materials by design：unconventional polymer networks give extreme properties. Chemical Reviews，2021，121（8）：4309-4372.

[5] Thomson R. Ductile versus brittle behaviour of crystalst. Philos Mag A，1974，29：73-97.

[6] McMeeking R，Evans A. Mechanics of transformation-toughening in brittle materials. Journal of the American Ceramic Society，1982，65（5）：242-246.

[7] Mallick P K. Fiber-reinforced composites: materials, manufacturing, and design. CRC press, 2007.

[8] Argon A S. The physics of deformation and fracture of polymers. New York: Cambridge, 2013.

[9] Fung Y C. Biomechanics: mechanical properties of living tissues. Springer Science & Business Media, 2013.

[10] Gong J P. Why are double network hydrogels so tough? Soft Matter, 2010, 6 (12): 2583-2590.

[11] Zhang T, Lin S, Yuk H, et al. Predicting fracture energies and crack-tip fields of soft tough materials. Extreme Mechanics Letters, 2015, 4: 1-8.

[12] Sun J Y, Zhao X, Illeperuma W R, et al. Highly stretchable and tough hydrogels. Nature, 2012, 489 (7414): 133-136.

[13] Yu Q M, Tanaka Y, Furukawa H, et al. Direct observation of damage zone around crack tips in double-network gels. Macromolecules 2009, 42 (12): 3852-3855.

[14] Gong J P, Katsuyama Y, Kurokawa T, et al. Double-network hydrogels with extremely high mechanical strength. Advanced materials, 2003, 15 (14): 1155-1158.

[15] Peppas N A, Bures P, Leobandung W, et al. Hydrogels in pharmaceutical formulations. European journal of pharmaceutics and biopharmaceutics, 2000, 50 (1): 27-46.

[16] Yang F, Tadepalli V, Wiley B J. 3D printing of a double network hydrogel with a compression strength and elastic modulus greater than those of cartilage. ACS Biomaterials Science & Engineering, 2017, 3 (5): 863-869.

[17] Ducrot E, Chen Y, Bulters M, et al. Toughening elastomers with sacrificial bonds and watching them break. Science, 2014, 344 (6180): 186-189.

[18] Du J, Xu S, Feng S, et al. Tough dual nanocomposite hydrogels with inorganic hybrid crosslinking. Soft Matter, 2016, 12 (6): 1649-1654.

[19] Chen Y, Shull K R. High-toughness polycation cross-linked triblock copolymer hydrogels. Macromolecules, 2017, 50 (9): 3637-3646.

[20] He Q, Huang Y, Wang S. Hofmeister effect-assisted one step fabrication of ductile and strong gelatin hydrogels. Advanced Functional Materials, 2018, 28 (5): 1705069.

[21] Sun Y N, Gao G R, Du G L, et al. Super tough, ultrastretchable, and thermoresponsive hydrogels with functionalized triblock copolymer micelles as macro-cross-linkers. ACS Macro Letters, 2014, 3 (5): 496-500.

[22] Zhong M, Liu X Y, Shi F K, et al. Self-healable, tough and highly stretchable ionic nanocomposite physical hydrogels. Soft Matter, 2015, 11 (21): 4235-4241.

[23] Foster E M, Lensmeyer E E, Zhang B, et al. Effect of polymer network architecture, enhancing soft materials using orthogonal dynamic bonds in an interpenetrating network. ACS Macro Letters, 2017, 6 (5): 495-499.

[24] Cao L, Fan J, Huang J, et al. A robust and stretchable cross-linked rubber network with recyclable and self-healable capabilities based on dynamic covalent bonds. Journal of Materials Chemistry A, 2019, 7 (9): 4922-4933.

[25] Fu J. Strong and tough hydrogels crosslinked by multi-functional polymer colloids. Journal of Polymer Science Part B: Polymer Physics, 2018, 56 (19): 1336-1350.

[26] Nakayama A, Kakugo A, Gong J P, et al. High mechanical strength double-network hydrogel with

bacterial cellulose. Advanced functional materials, 2004, 14 (11): 1124-1128.

[27] Toivonen M S, Kurki-Suonio S, Schacher F H, et al. Water-resistant, transparent hybrid nanopaper by physical cross-linking with chitosan. Biomacromolecules, 2015, 16 (3): 1062-1071.

[28] Ye D, Cheng Q, Zhang Q, et al. Deformation drives alignment of nanofibers in framework for inducing anisotropic cellulose hydrogels with high toughness. ACS applied materials & interfaces, 2017, 9 (49): 43154-43162.

[29] Brown A E, Litvinov R I, Discher D E, et al. Multiscale mechanics of fibrin polymer: gel stretching with protein unfolding and loss of water. Science, 2009, 325 (5941): 741-744.

[30] Wang W, Zhang Y, Liu W. Bioinspired fabrication of high strength hydrogels from non-covalent interactions. Progress in Polymer Science, 2017, 71: 1-25.

[31] Dai X, Zhang Y, Gao L, et al. A Mechanically Strong, Highly Stable, Thermoplastic, and self-healable supramolecular polymer hydrogel. Advanced Materials, 2015, 27 (23): 3566-3571.

[32] Zhang Y, Li Y, Liu W. Dipole–dipole and H-bonding interactions significantly enhance the multifaceted mechanical properties of thermoresponsive shape memory hydrogels. Advanced Functional Materials, 2015, 25 (3): 471-480.

[33] Nakahata M, Takashima Y, Harada A. Highly flexible, tough, and self-healing supramolecular polymeric materials using host–guest interaction. Macromolecular rapid communications, 2016, 37 (1): 86-92.

[34] Gonzalez M A, Simon J R, Ghoorchian A, et al. strong, tough stretchable, and self-adhesive hydrogels from intrinsically unstructured proteins. Advanced Materials, 2017, 29 (10): 1604743.

[35] Mayumi K, Guo J, Narita T, et al. Fracture of dual crosslink gels with permanent and transient crosslinks. Extreme Mechanics Letters, 2016, 6: 52-59.

[36] Zhang X N, Wang Y J, Sun S, et al. A tough and stiff hydrogel with tunable water content and mechanical properties based on the synergistic effect of hydrogen bonding and hydrophobic interaction. Macromolecules, 2018, 51 (20): 8136-8146.

[37] Zhang Y, Yong Y, An D, et al. A drip-crosslinked tough hydrogel. Polymer, 2018, 135: 327-330.

[38] Takashima Y, Sawa Y, Iwaso K, et al. Supramolecular materials cross-linked by host–guest inclusion complexes: the effect of side chain molecules on mechanical properties. Macromolecules, 2017, 50 (8): 3254-3261.

[39] Zheng S Y, Ding H, Qian J, et al. Metal-coordination complexes mediated physical hydrogels with high toughness, stick–slip tearing behavior, and good processability. Macromolecules, 2016, 49 (24): 9637-9646.

[40] Luo F, Sun T L, Nakajima T, et al. Oppositely charged polyelectrolytes form tough, self-healing, and rebuildable hydrogels. Advanced materials, 2015, 27 (17): 2722-2727.

[41] Yang Y, Wang X, Yang F, et al. Highly elastic and ultratough hybrid ionic–covalent hydrogels with tunable structures and mechanics. Advanced Materials, 2018, 30 (18): 1707071.

[42] Bakarich S E, Gorkin Ⅲ R, Panhuis M I H, et al. 4D printing with mechanically robust, thermally actuating hydrogels. Macromolecular rapid communications, 2015, 36 (12): 1211-1217.

[43] Li J, Illeperuma W R, Suo Z, et al. Hybrid hydrogels with extremely high stiffness and toughness. ACS Macro Letters, 2014, 3 (6): 520-523.

[44] Hu X, Vatankhah-Varnoosfaderani M, Zhou J, et al. Weak hydrogen bonding enables hard, strong, tough, and elastic hydrogels. Advanced materials, 2015, 27 (43): 6899-6905.

[45] Zheng W J, An N, Yang J H, et al. Tough Al-alginate/poly (N-isopropylacrylamide) hydrogel with tunable LCST for soft robotics. ACS applied materials & interfaces, 2015, 7 (3): 1758-1764.

[46] Li C, Rowland M J, Shao Y, et al. Responsive double network hydrogels of interpenetrating DNA and CB [8] host–guest supramolecular systems. Advanced materials, 2015, 27 (21): 3298-3304.

[47] Chen Q, Yan X, Zhu L, et al. Improvement of mechanical strength and fatigue resistance of double network hydrogels by ionic coordination interactions. Chemistry of Materials, 2016, 28 (16): 5710-5720.

[48] Jia H, Huang Z, Fei Z, et al. Unconventional tough double-network hydrogels with rapid mechanical recovery, self-healing, and self-gluing properties. ACS applied materials & interfaces, 2016, 8 (45): 31339-31347.

[49] Chen Q, Chen H, Zhu L, et al. Engineering of tough double network hydrogels. Macromolecular Chemistry and Physics, 2016, 217 (9): 1022-1036.

[50] Guo M, Pitet L M, Wyss H M, et al. Tough stimuli-responsive supramolecular hydrogels with hydrogen-bonding network junctions. Journal of the American Chemical Society, 2014, 136 (19): 6969-6977.

[51] Shao C, Chang H, Wang M, et al. High-strength, tough, and self-healing nanocomposite physical hydrogels based on the synergistic effects of dynamic hydrogen bond and dual coordination bonds. ACS applied materials & interfaces, 2017, 9 (34): 28305-28318.

[52] Gao G, Du G, Sun Y, et al. Self-healable, tough, and ultrastretchable nanocomposite hydrogels based on reversible polyacrylamide/montmorillonite adsorption. ACS applied materials & interfaces, 2015, 7 (8): 5029-5037.

[53] Yang J, Han C R, Zhang X M, et al. Cellulose nanocrystals mechanical reinforcement in composite hydrogels with multiple cross-links: correlations between dissipation properties and deformation mechanisms. Macromolecules, 2014, 47 (12): 4077-4086.

[54] Cui K, Sun T L, Liang X, et al. Multiscale energy dissipation mechanism in tough and self-healing hydrogels. Physical review letters, 2018, 121 (18): 185501.

[55] Grandbois M, Beyer M, Rief M, et al. How strong is a covalent bond? Science, 1999, 283 (5408): 1727-1730.

[56] Zhao X. Designing toughness and strength for soft materials. Proceedings of the National Academy of Sciences, 2017, 114 (31): 8138-8140.

[57] Griffith A A. The phenomena of rupture and flow in solids. Philosophical transactions of the royal society of london Series A, containing papers of a mathematical or physical character, 1921, 221 (582-593): 163-198.

[58] Gao H, Ji B, Jäger I L, et al. Materials become insensitive to flaws at nanoscale: lessons from nature. Proceedings of the national Academy of Sciences, 2003, 100 (10): 5597-5600.

[59] Chen C, Wang Z, Suo Z. Flaw sensitivity of highly stretchable materials. Extreme Mechanics Letters, 2017, 10: 50-57.

[60] Lin S, Liu X, Liu J, et al. Anti-fatigue-fracture hydrogels. Science advances, 2019, 5 (1):

eaau8528.

[61] Lin S, Liu J, Liu X, et al. Muscle-like fatigue-resistant hydrogels by mechanical training. Proceedings of the National Academy of Sciences, 2019, 116 (21): 10244-10249.

[62] Keten S, Xu Z, Ihle B, et al. Nanoconfinement controls stiffness, strength and mechanical toughness of β-sheet crystals in silk. Nature materials, 2010, 9 (4): 359-367.

[63] Liao I C, Moutos F T, Estes B T, et al. Composite three-dimensional woven scaffolds with interpenetrating network hydrogels to create functional synthetic articular cartilage. Advanced functional materials, 2013, 23 (47): 5833-5839.

[64] Lin S, Cao C, Wang Q, et al. Design of stiff, tough and stretchy hydrogel composites via nanoscale hybrid crosslinking and macroscale fiber reinforcement. Soft matter, 2014, 10 (38): 7519-7527.

[65] Illeperuma W R, Sun J Y, Suo Z, et al. Fiber-reinforced tough hydrogels. Extreme Mechanics Letters, 2014, 1: 90-96.

[66] Agrawal A, Rahbar N, Calvert P D. Strong fiber-reinforced hydrogel. Acta biomaterialia, 2013, 9 (2): 5313-5318.

[67] King D R, Sun T. L, Huang Y, et al. Extremely tough composites from fabric reinforced polyampholyte hydrogels. Materials horizons, 2015, 2 (6): 584-591.

[68] Kong W, Wang C, Jia C, et al. Muscle-inspired highly anisotropic, strong, ion-conductive hydrogels. Advanced Materials, 2018, 30 (39): 1801934.

[69] Gent A N, Walter J D. Pneumatic tire. 2006.

[70] Lin S, Zhou Y, Zhao X. Designing extremely resilient and tough hydrogels via delayed dissipation. Extreme Mechanics Letters, 2014, 1: 70-75.

[71] Cui J, Lackey M A, Madkour A E, et al. Synthetically simple, highly resilient hydrogels. Biomacromolecules, 2012, 13 (3): 584-588.

[72] Kamata H, Akagi Y, Kayasuga-Kariya Y, et al. "Nonswellable" hydrogel without mechanical hysteresis. Science, 2014, 343 (6173): 873-875.

[73] Sacks M S, Yoganathan A P. Heart valve function: a biomechanical perspective. Philosophical Transactions of the Royal Society B: Biological Sciences, 2007, 362 (1484): 1369-1391.

[74] Sacks M S, Merryman W D, Schmidt D E., On the biomechanics of heart valve function. Journal of biomechanics, 2009, 42 (12): 1804-1824.

[75] Le Cam J B. Energy storage due to strain-induced crystallization in natural rubber: The physical origin of the mechanical hysteresis. Polymer, 2017, 127: 166-173.

[76] Xiang C, Wang Z, Yang C, et al. Stretchable and fatigue-resistant materials. Materials Today, 2020, 34: 7-16.

[77] Li J, Celiz A, Yang J, et al. Tough adhesives for diverse wet surfaces. Science, 2017, 357 (6349): 378-381.

[78] Volinsky A, Moody N, Gerberich W. Interfacial toughness measurements for thin films on substrates. Acta materialia, 2002, 50 (3): 441-466.

[79] Yuk H, Zhang T, Lin S, et al. Tough bonding of hydrogels to diverse non-porous surfaces. Nature materials, 2016, 15 (2): 190-196.

[80] Rose S, Prevoteau A, Elzière P, et al. Nanoparticle solutions as adhesives for gels and biological

tissues. Nature, 2014, 505 (7483): 382-385.
[81] Yuk H, Zhang T, Parada G A, et al.Skin-inspired hydrogel–elastomer hybrids with robust interfaces and functional microstructures. Nature communications, 2016, 7 (1): 12028.
[82] Zhang T, Yuk H, Lin S, et al. Tough and tunable adhesion of hydrogels: experiments and models. Acta Mechanica Sinica, 2017, 33: 543-554.
[83] Liu J, Lin S, Liu X, et al. Fatigue-resistant adhesion of hydrogels. Nature communications, 2020, 11 (1): 1071.
[84] Lee H, Dellatore S M, Miller W M, et al. Mussel-inspired surface chemistry for multifunctional coatings. Science, 2007, 318 (5849): 426-430.
[85] Yuk H, Lin S, Ma C, et al. Hydraulic hydrogel actuators and robots optically and sonically camouflaged in water. Nature communications, 2017, 8 (1): 14230.
[86] Wirthl D, Pichler R, Drack M, et al. Instant tough bonding of hydrogels for soft machines and electronics. Science advances, 2017, 3 (6): e1700053.
[87] Yang J, Bai R, Suo Z. Topological adhesion of wet materials. Advanced Materials, 2018, 30 (25): 1800671.
[88] Kurokawa T, Furukawa H, Wang W, et al. Formation of a strong hydrogel–porous solid interface via the double-network principle. Acta biomaterialia, 2010, 6 (4): 1353-1359.
[89] Rao P, Sun T L, Chen L, et al. Tough hydrogels with fast, strong, and reversible underwater adhesion based on a multiscale design. Advanced Materials, 2018, 30 (32): 1801884.
[90] Yuk H, Varela C E, Nabzdyk C S, et al. Dry double-sided tape for adhesion of wet tissues and devices. Nature, 2019, 575 (7781): 169-174.
[91] Yang J, Bai R, Chen B, et al. Hydrogel adhesion: a supramolecular synergy of chemistry, topology, and mechanics. Advanced Functional Materials, 2020, 30 (2): 1901693.
[92] Ghobril C, Grinstaff M. The chemistry and engineering of polymeric hydrogel adhesives for wound closure: a tutorial. Chemical Society Reviews, 2015, 44 (7): 1820-1835.
[93] Pizzi A, Mittal K L. Handbook of adhesive technology. CRC press, 2017.
[94] Evans A, Hutchinson J, Wei Y. Interface adhesion: effects of plasticity and segregation. Acta materialia, 1999, 47 (15-16): 4093-4113.
[95] Raphael E, De Gennes P. Rubber-rubber adhesion with connector molecules. The Journal of Physical Chemistry, 1992, 96 (10): 4002-4007.
[96] Gent A, Petrich R. Adhesion of viscoelastic materials to rigid substrates. Proceedings of the Royal Society of London A Mathematical and Physical Sciences, 1969, 310 (1502): 433-448.
[97] Walia R, Akhavan B, Kosobrodova E, et al. Hydrogel-solid hybrid materials for biomedical applications enabled by surface-embedded radicals. Advanced Functional Materials, 2020, 30 (38): 2004599.
[98] Mao S, Zhang D, Zhang Y, et al. A universal coating strategy for controllable functionalized polymer surfaces. Advanced Functional Materials, 2020, 30 (40): 2004633.
[99] Wei K, Chen X, Zhao P, et al. Stretchable and bioadhesive supramolecular hydrogels activated by a one-stone–two-bird postgelation functionalization method. ACS applied materials & interfaces, 2019, 11 (18): 16328-16335.

[100] Liu Q, Nian G, Yang C, et al. Bonding dissimilar polymer networks in various manufacturing processes. Nature communications, 2018, 9 (1): 846.

[101] Kord Forooshani P, Lee B P. Recent approaches in designing bioadhesive materials inspired by mussel adhesive protein. Journal of Polymer Science Part A: Polymer Chemistry, 2017, 55 (1): 9-33.

[102] Saiz-Poseu J, Mancebo-Aracil J, Nador F, et al. The chemistry behind catechol-based adhesion. Angewandte Chemie International Edition, 2019, 58 (3): 696-714.

[103] Chen H, Liu Y, Ren B, et al. Super bulk and interfacial toughness of physically crosslinked double-network hydrogels. Advanced Functional Materials, 2017, 27 (44): 1703086.

[104] Liu J, Tan C S Y, Scherman O A. Dynamic interfacial adhesion through cucurbit [n] uril molecular recognition. Angewandte Chemie, 2018, 130 (29): 8992-8996.

[105] Gao Y, Wu K, Suo Z. Photodetachable adhesion. Advanced Materials, 2019, 31 (6): 1806948.

[106] Liu X, Zhang Q, Duan L, et al. Tough Adhesion of Nucleobase-Tackifed Gels in Diverse Solvents. Advanced Functional Materials, 2019, 29 (17): 1900450.

[107] Fan H, Wang J, Tao Z, et al. Adjacent cationic–aromatic sequences yield strong electrostatic adhesion of hydrogels in seawater. Nature communications, 2019, 10 (1): 5127.

[108] Gao Y, Chen J, Han X, et al. Universal strategy for tough adhesion of wet soft material. Advanced Functional Materials, 2020, 30 (36): 2003207.

[109] Tian K, Bae, J, Suo Z, et al. Adhesion between hydrophobic elastomer and hydrogel through hydrophilic modification and interfacial segregation. ACS applied materials & interfaces, 2018, 10 (49): 43252-43261.

[110] Ji H, De Gennes P. Adhesion via connector molecules: the many-stitch problem. Macromolecules, 1993, 26 (3): 520-525.

[111] Tamesue S, Endo T, Ueno Y, et al. Sewing hydrogels: adhesion of hydrogels utilizing in situ polymerization of linear polymers inside gel networks. Macromolecules, 2019, 52 (15): 5690-5697.

[112] Takahashi R, Shimano K, Okazaki H, et al. Tough particle-based double network hydrogels for functional solid surface coatings. Advanced Materials Interfaces, 2018, 5 (23): 1801018.

[113] Cho H, Wu G, Christopher Jolly J, et al. Intrinsically reversible superglues via shape adaptation inspired by snail epiphragm. Proceedings of the National Academy of Sciences, 2019, 116 (28): 13774-13779.

[114] Yang S Y, O' Cearbhaill E D, Sisk G C, et al. A bio-inspired swellable microneedle adhesive for mechanical interlocking with tissue. Nature communications, 2013, 4 (1): 1702.

[115] Fullenkamp D E, He L, Barrett D G, Burghardt W R, et al. Mussel-inspired histidine-based transient network metal coordination hydrogels. Macromolecules, 2013, 46 (3): 1167-1174.

[116] Ahn B K, Das S, Linstadt R, et al. High-performance mussel-inspired adhesives of reduced complexity. Nature communications, 2015, 6 (1): 8663.

[117] Zhang W, Wang R, Sun Z, et al. Catechol-functionalized hydrogels: biomimetic design, adhesion mechanism, and biomedical applications. Chemical Society Reviews, 2020, 49 (2): 433-464.

[118] Bai R, Yang J, Suo Z. Fatigue of hydrogels. European Journal of Mechanics-A/Solids, 2019, 74: 337-370.

[119] Bai, R, Yang J, Morelle X P, et al. Flaw-insensitive hydrogels under static and cyclic loads. Macromolecular rapid communications, 2019, 40 (8): 1800883.

[120] Baker M I, Walsh S P, Schwartz Z, et al. A review of polyvinyl alcohol and its uses in cartilage and orthopedic applications. Journal of Biomedical Materials Research Part B: Applied Biomaterials, 2012, 100 (5): 1451-1457.

[121] Lake G, Lindley P. The mechanical fatigue limit for rubber. Journal of Applied Polymer Science, 1965, 9 (4): 1233-1251.

[122] Lake G, Thomas A. The strength of highly elastic materials. Proceedings of the Royal Society of London Series A Mathematical and Physical Sciences, 1967, 300 (1460): 108-119.

[123] Bai R, Yang Q, Tang J, et al. Fatigue fracture of tough hydrogels. Extreme Mechanics Letters, 2017, 15: 91-96.

[124] Li X, Cui K, Sun T L, et al. Mesoscale bicontinuous networks in self-healing hydrogels delay fatigue fracture. Proceedings of the National Academy of Sciences, 2020, 117 (14): 7606-7612.

[125] Wang Z, Xiang C, Yao X, et al. Stretchable materials of high toughness and low hysteresis. Proceedings of the National Academy of Sciences, 2019, 116 (13): 5967-5972.

[126] Ni X, Chen C, Li J. Interfacial fatigue fracture of tissue adhesive hydrogels. Extreme Mechanics Letters, 2020, 34: 100601.

[127] Zhang W, Gao Y, Yang H, et al. Fatigue-resistant adhesion I. Long-chain polymers as elastic dissipaters. Extreme Mechanics Letters, 2020, 39: 100813.

[128] Baumard T, Thomas A, Busfield J. Fatigue peeling at rubber interfaces. Plastics, rubber and composites, 2012, 41 (7): 296-300.

[129] Ushio K, Oka M, Hyon S H, et al. Attachment of artificial cartilage to underlying bone. Journal of Biomedical Materials Research Part B: Applied Biomaterials: An Official Journal of The Society for Biomaterials, The Japanese Society for Biomaterials, and The Australian Society for Biomaterials and the Korean Society for Biomaterials, 2004, 68 (1): 59-68.

第 7 章
水凝胶功能特性的设计原理和调控方法

7.1 导电性：形成连通的电子导电相
7.2 磁性：嵌入磁性颗粒并形成铁磁磁畴
7.3 折射率和透明度：均匀嵌入高折射率且无散射的纳米相
7.4 可调控声阻抗：等效均质水凝胶的密度和体积模量的调控
7.5 自愈性：在损伤区域形成新的交联或高分子链
7.6 可注射性：选择具有剪切变稀和自我修复特性的材料
参考文献

除具有第 6 章中讨论的极端力学性能外，具有其他优异物理特性的水凝胶的设计在近年来吸引了越来越多的关注。目前正在开发和探索的水凝胶的物理特性包括高导电性[1]、图案化磁化[2]、高折射率和透明度[3,4]、可调谐声阻抗[5]、自我修复[6] 和可注射等。这些优异的物理特性与水凝胶的各种应用的极端力学性能具有相似的重要性，尤其是对于水凝胶器件[7]。在本章中，我们将讨论具有这些物理特性的水凝胶的设计原则和实现策略。

7.1 导电性：形成连通的电子导电相

导电性对于水凝胶的创新应用至关重要。例如用于刺激和记录生物电子学中神经活动的水凝胶生物电极[1]，以及用于超级电容器和储能电池的水凝胶电极[7,8]。然而，普通水凝胶的电导率低于几西门子每米，与盐水处于同一水平[1]。与高导电性的金属、碳、导电高分子等相比，普通水凝胶的导电能力几乎可以忽略不计。

导电水凝胶的设计原理是将液态金属、金属纳米线、碳纳米管、石墨烯、导电聚合物等导电相嵌入水凝胶基质中，使导电相在水凝胶中形成渗透网络 [也被称为渗透导电相，图 7.1（a）][1,9,10]。特别是基于导电聚合物的导电水凝胶，由于其独特的聚合物性质、良好的电气性能、力学性能、稳定性以及生物相容性，引起了相关学者极大的研究兴趣[11-18]。例如，聚 3,4- 乙烯二氧噻吩：聚苯乙烯磺酸

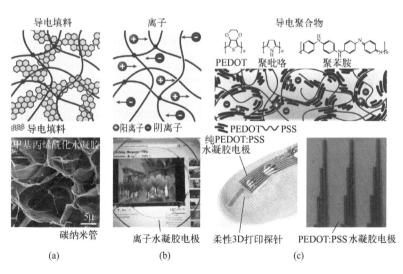

图 7.1　高导电性水凝胶的设计原理：渗透导电相
（a）具有渗透导电填料的水凝胶[10]；（b）具有离子导电盐溶剂的水凝胶[19]；
（c）基于导电聚合物的水凝胶[16,18]

盐（PEDOT ∶ PSS）已被制成纯导电聚合物水凝胶，可实现每米几千西门子的高电导率和卓越的生物相容性 [图 7.1（c）][16-18]。除了上述导电水凝胶，基于离子导电的水凝胶进一步被广泛开发为可拉伸和透明的离子导体，从而用于各种生物医学应用 [图 7.1（b）][19]。离子导电水凝胶中的导电相通常是高浓度的盐离子[1]。

7.2 磁性：嵌入磁性颗粒并形成铁磁磁畴

具有铁磁畴或磁化的弹性体和水凝胶等软材料由于它们的机械顺应性，潜在的生物相容性和在施加磁场下快速变形的能力，已被广泛开发并用于药物输送和微创手术等生物医学领域 [2,20-25]。普通水凝胶通常是抗磁性的，不包含铁磁域，具有与水相似的磁性。因此，在外加磁场作用下，普通水凝胶无法变形，施加力或释放物质。

水凝胶具有图案化磁化的设计原理是：在水凝胶基质中嵌入硬磁、软磁或超顺磁颗粒等磁性成分，其中的铁磁畴可以进一步图案化。换言之，水凝胶图案化磁化的核心点是嵌入磁性颗粒，然后图案化相应的铁磁畴（图 7.2）[2,20-23,25]。此外，由于硬磁颗粒的高矫顽力，磁饱和后硬磁颗粒（钕铁硼颗粒）可以在驱动磁场中保持其磁化强度 [图 7.2（a）]。因此，可以将图案化的铁磁畴编程到嵌有硬磁颗粒的弹性体和水凝胶中。当受到驱动磁场的影响后，具有图案化铁磁畴的弹性体和水凝胶可以在各种形状之间快速转换 [20-23,25]。最近，3D 打印已成为铁磁域图案化的一种有效的方法，可用于编程铁磁弹性体和水凝胶中的复杂 3D 形状以及磁畴图案 [图 7.2（b）、（c）][20,23]。 值得注意的是，磁性颗粒在水凝胶基质的水性环境中可能具有腐蚀性。为了增强它们在水凝胶基质中的化学稳定性，通常会给磁性颗粒涂上一层保护层（二氧化硅等）[图 7.2（b）、（c）][25]。

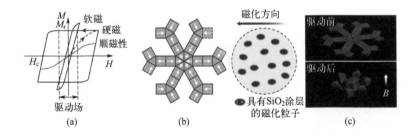

图 7.2 具有图案化磁化的水凝胶和弹性体的设计原理：嵌入磁性颗粒和图案化铁磁畴
（a）顺磁、软磁和硬磁材料的外加磁场 H 和磁化强度 M 之间的典型关系，M_r 和 H_c 分别是硬磁材料的剩余磁化强度和矫顽力[26]；（b）硬磁颗粒可以嵌入弹性体/水凝胶基质中，其中的铁磁域可以通过 3D 打印进行图案化[27]；（c）所得磁性软材料在磁性驱动前后的照片[27]

7.3 折射率和透明度：均匀嵌入高折射率且无散射的纳米相

水凝胶在眼科镜片[28-30]、光纤[31,32]等光学领域的应用，需要水凝胶具有高折射率和高透明度 [图 7.3（a）]。普通水凝胶的折射率与水相似，约为 1.333。提高水凝胶折射率的一种通用策略是在水凝胶基质中均匀嵌入纳米相，如纳米颗粒和具有高折射率的纳米晶域[3,4]。然而，纳米相和水凝胶基质之间的折射率不匹配可能会导致大量非必要的光散射，从而降低水凝胶的透明度 [图 7.3（b）]。研究发现，只要将纳米相的尺寸减小到光波长的 1/10 以下（如直径为 3nm 的硫化锌纳米颗粒）就可以有效地减少光散射，从而获得具有高折射率（约 1.49）和高透明度的水凝胶 [图 7.3（b）～（d）][3]。总而言之，具有高折射率和透明度的水凝胶的设计原则是将高折射率非散射尺寸远小于光波长的纳米相均匀地嵌入水凝胶基质中。

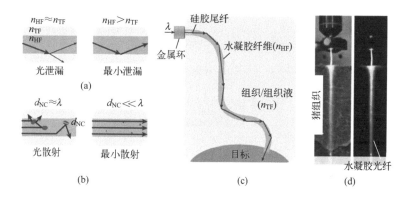

图 7.3 具有高折射率和高透明度的水凝胶的设计原则：均匀嵌入高折射率非散射纳米相[32]（a）水凝胶纤维（n_{HF}）和组织液（n_{TF}）的反射率之间的高对比度可以使光泄漏最小化；（b）在水凝胶基质中均匀嵌入纳米相（例如具有高折射率的纳米粒子），可以提高水凝胶的折射率，纳米相的尺寸（d_{NC}）应远小于光波长 λ，以实现最小散射和高透明度；（c）具有高反射率和高透明度的水凝胶用于活体组织中的光纤；（d）水凝胶光纤的照片

7.4 可调控声阻抗：等效均质水凝胶的密度和体积模量的调控

水凝胶可作为一种有效的声波传输介质（偶联剂）用于成像和超声治疗。因此，需要设计具有可调声阻抗特性的水凝胶以匹配不同的材料或不同的环境

阻抗[5,33]。通常，均质材料的声阻抗 Z 可表示为：

$$Z = \sqrt{\rho_{\text{eff}} K_{\text{eff}}}$$

式中，ρ_{eff} 和 K_{eff} 分别是材料的有效密度和体积模量。

由于普通水凝胶的密度和体积模量与水几乎相同，因此普通水凝胶的声阻抗也与水近似。为了赋予水凝胶可调声阻抗的特性，研究人员在坚韧水凝胶基质中构建相应的流体通道 [图 7.4（a）][5]。通过将空气、水或液态金属（即共晶镓-铟）注入这些流体通道，水凝胶的有效密度、体积模量和声阻抗等参数可以根据需要调节成接近于空气、水或许多其他固体 [图 7.4（b）][5]。为了使水凝胶更加接近均质材料，流体通道应均匀分布在水凝胶中，且流体通道的特征尺寸（即通道直径和相邻通道之间的距离）应远小于声波波长。总而言之，具有可调声阻抗的水凝胶的通用设计原则是通过构建流体通道来调整有效均匀水凝胶的密度和体积模量。

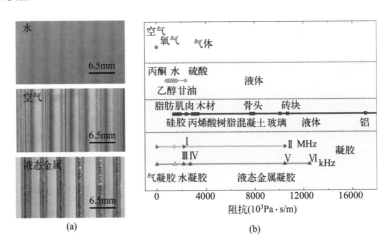

图 7.4　具有可调声阻抗的水凝胶的设计原则：调整有效均匀水凝胶的密度和体积模量[5]
（a）通过将空气、水或液态金属（即共晶镓-铟）注入水凝胶基质内的流体通道中，可以显著改变水凝胶的有效密度、体积模量和声阻抗；（b）水凝胶可以根据需要实现近似空气、水或许多固体的声阻抗

7.5　自愈性：在损伤区域形成新的交联或高分子链

许多生物水凝胶的一个显著特征是它们在受伤后具有愈合能力。这种自修复能力可以赋予合成水凝胶以减轻损伤和长期稳健等优点。然而，生物水凝胶的愈

合过程主要依赖于生物细胞的功能，而合成水凝胶通常不存在这些功能。在没有活性成分的情况下，实现工程材料自愈性的通用策略是在受损区域附近形成新的材料和/或相互作用[34]。

对于弹性体和水凝胶等软材料而言，在受损区域附近形成新材料通常是指形成新的交联和/或聚合物链，这也是自修复水凝胶的设计原则[图7.5（a）]。用于水凝胶自修复的常用交联包括氢键[35-38]、离子键[39-41]、金属配位[13,37,41]、疏水相互作用[42]、主客体相互作用[43]以及动态共价键[44,45]等弱物理相互作用。一旦受损水凝胶中两个新形成的表面在特定温度和pH等特定条件下相互接触，界面上就会形成新的交联，从而赋予水凝胶自愈能力[图7.5（a）、（c）]。除了弱物理相互作用和动态共价交联（即可逆交联）之外，水凝胶的自愈还可以通过聚合物链的相互扩散形成跨越裂缝表面的链缠结来实现[图7.5（a）、（c）][46,47]。

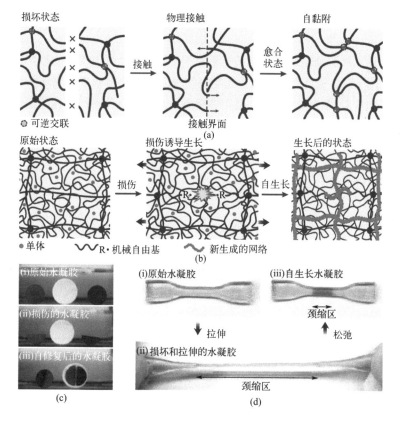

图7.5 自修复水凝胶的设计原则：在受损区域形成交联和/或聚合物链

（a）可逆交联和聚合物链缠结在两片水凝胶之间的界面上形成，用于自修复或自黏附；（b）水凝胶的损伤会引发新的聚合和交联，从而使水凝胶自增强或自生长[49]；（c）基于带相反电荷的聚电解质的自修复水凝胶的照片[40, 50]；（d）自增强/自生长水凝胶的照片[49]

因此，水凝胶的自愈过程主要涉及链缠结和可逆交联两种机制[48]。值得注意的是，当两个完整的自修复水凝胶的表面在特定温度、pH等条件下接触时，也可以在两个水凝胶之间形成黏附界面。严格来说，现有的自修复水凝胶大多是自粘水凝胶，因为水凝胶的损伤不需要诱导自修复过程。Matsuda等报道了一种自生长/自增强水凝胶，水凝胶中聚合物链的断裂可以诱导机械自由基的出现，进而触发单体在水凝胶溶剂中的聚合[图7.5（b）][49]。由于水凝胶在中度损伤后可以实现类似于人体肌肉的机械训练的自生长或自增强[图7.5（d）]，该策略未来可以用于真正自修复（而不是自粘）水凝胶的设计，其愈合能力是由损伤触发的[6]。

7.6 可注射性：选择具有剪切变稀和自我修复特性的材料

可注射水凝胶使微创策略能够在不需手术植入的情况下提供治疗药物和细胞的输送（图7.6）。虽然水凝胶的流变学特性受到直接注射或导管输送药物需求的限制，但水凝胶在临床环境中的适用性仍然很高，从身体不同区域的定位到靶向输送各种目标物等。在这里，我们关注并讨论现有可注射水凝胶的流变学特性，并强调确定特性-功能关系以促进其临床应用的必要性。

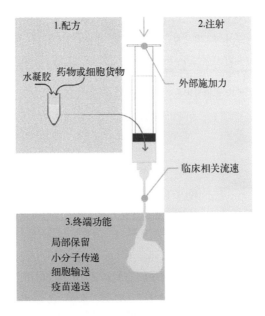

图7.6 可注射水凝胶必须适应配方，注射和终端功能的限制[51]

流动的材料必须表现出类似液体的行为，其组成分子就能够在相关的加工条件下相互移动。然而，大多数共价材料由于其共价键阻止了组成分子的相对运动，并不具有流动性，因此，"静态"共价水凝胶需要注入在注射时可以实现凝胶化的预聚物体系或可以响应温度、紫外线、pH等外部刺激而实现交联的刺激响应聚合物。最近，越来越多的研究开始将动态交联水凝胶作为可注射材料[52-55]。尽管具体的交联机制各不相同，但通常这些方法均能使水凝胶具有动态、屈服和自修复的流变响应特性。通常，物理交联会在静态条件下形成具有类似固体材料特性的水凝胶网络。然而，当变形时，动态交联会被破坏，从而分散应力并导致类似液体的行为；由于交联是可逆的，它们可以在变形后重新结合以恢复网络结构及其类似固体的特性（图7.7）。

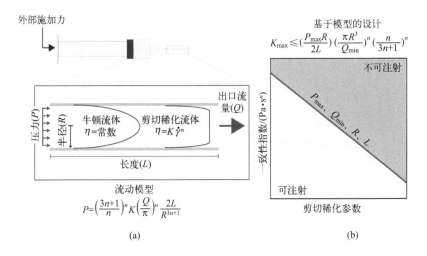

图7.7 （a）流动模型与注射压力（P）、流速（Q）、黏度参数（K、n）和几何形状（R、L）相关，允许在各种临床情况下计算注射压力。（b）反向使用流动模型来识别容易满足所需过程和几何约束的材料参数。满足条件的参数组合（K和n）显示在Ashby样式图中，其中线上方的组合将导致无法达到的压力或流速太慢[51]

参考文献

[1] Yuk H，Lu B，Zhao X. Hydrogel bioelectronics. Chemical Society Reviews，2019，48（6）：1642-1667.

[2] Zrinyi M，Barsi L，Büki A. Ferrogel：a new magneto-controlled elastic medium. Polymer Gels and Networks，1997，5（5）：415-427.

[3] Zhang Q，Fang Z，Cao Y，et al. High refractive index inorganic–organic interpenetrating polymer network（IPN）hydrogel nanocomposite toward artificial cornea implants. ACS Macro Letters，2012，1

(7): 876-881.
[4] Lü C, Yang B. High refractive index organic–inorganic nanocomposites: design, synthesis and application. Journal of Materials Chemistry, 2009, 19 (19): 2884-2901.
[5] Zhang K, Ma C, He Q, et al. Metagel with broadband tunable acoustic properties over air–water–solid ranges. Advanced Functional Materials, 2019, 29 (38): 1903699.
[6] Taylor D L, in het Panhuis M. Self-healing hydrogels. Advanced Materials, 2016, 28 (41): 9060-9093.
[7] Liu X, Liu J, Lin S, et al. Hydrogel machines. Materials Today, 2020, 36: 102-124.
[8] Guo Y, Bae J, Fang Z, et al. Hydrogels and hydrogel-derived materials for energy and water sustainability. Chemical reviews, 2020, 120 (15): 7642-7707.
[9] Takahashi R, Sun T L, Saruwatari Y, et al. Creating stiff, tough, and functional hydrogel composites with low-melting-point alloys. Advanced Materials, 2018, 30 (16): 1706885.
[10] Shin S R, Jung S M, Zalabany M, et al. Carbon-nanotube-embedded hydrogel sheets for engineering cardiac constructs and bioactuators. ACS nano, 2013, 7 (3): 2369-2380.
[11] Zhang S, Chen Y, Liu H, et al. Room-temperature-formed PEDOT: PSS hydrogels enable injectable, soft, and healable organic bioelectronics. Advanced Materials, 2020, 32 (1): 1904752.
[12] Zhao X, Yuk H, Inoue A. Strong adhesion of conducting polymers on diverse substrates, Google Patents, 2020.
[13] Shi Y, Wang M, Ma C, et al. A conductive self-healing hybrid gel enabled by metal–ligand supramolecule and nanostructured conductive polymer. Nano letters, 2015, 15 (9): 6276-6281.
[14] Shi Y, Ma C, Peng L, et al. Conductive "smart" hybrid hydrogels with PNIPAM and nanostructured conductive polymers. Advanced Functional Materials, 2015, 25 (8): 1219-1225.
[15] Feig V R, Tran H, Lee M, et al. Mechanically tunable conductive interpenetrating network hydrogels that mimic the elastic moduli of biological tissue. Nature communications, 2018, 9 (1): 2740.
[16] Liu Y, Liu J, Chen S, et al. Soft and elastic hydrogel-based microelectronics for localized low-voltage neuromodulation. Nature biomedical engineering, 2019, 3 (1): 58-68.
[17] Lu B, Yuk H, Lin S, et al. Pure pedot: Pss hydrogels. Nature communications, 2019, 10 (1): 1043.
[18] Yuk H, Lu B, Lin S, et al. 3D printing of conducting polymers. Nature communications, 2020, 11 (1): 1604.
[19] Suo Z. Stretchable, transparent, ionic conductors. Science, 341 (6149): 984.
[20] Jeon S, Hoshiar A K, Kim K, et al. A magnetically controlled soft microrobot steering a guidewire in a three-dimensional phantom vascular network. Soft robotics, 2019, 6 (1): 54-68.
[21] Edelmann J, Petruska A J, Nelson B J. Magnetic control of continuum devices. The International Journal of Robotics Research, 2017, 36 (1): 68-85.
[22] Hu W, Lum G Z, Mastrangeli M, et al. Small-scale soft-bodied robot with multimodal locomotion. Nature, 2018, 554 (7690): 81-85.
[23] Xu T, Zhang J, Salehizadeh M, et al. Millimeter-scale flexible robots with programmable three-dimensional magnetization and motions. Science Robotics, 2019, 4 (29): eaav4494.
[24] Zhao X, Kim J, Cezar C A, et al. Active scaffolds for on-demand drug and cell delivery. Proceedings

of the National Academy of Sciences, 2011, 108 (1): 67-72.

[25] Kim Y, Parada G A, Liu S, et al. Ferromagnetic soft continuum robots. Science Robotics, 2019, 4 (33): eaax7329.

[26] Wang L, Kim Y, Guo C F, et al. Hard-magnetic elastica. Journal of the Mechanics and Physics of Solids, 2020, 142: 104045.

[27] Kim Y, Yuk H, Zhao R, et al. Printing ferromagnetic domains for untethered fast-transforming soft materials. Nature, 2018, 558 (7709): 274-279.

[28] Dong L, Agarwal A K, Beebe D J, et al. Adaptive liquid microlenses activated by stimuli-responsive hydrogels. Nature, 2006, 442 (7102): 551-554.

[29] Nicolson P C, Vogt J. Soft contact lens polymers: an evolution. Biomaterials, 2001, 22 (24): 3273-3283.

[30] De Groot J H, van Beijma F J, Haitjema H J, et al. Injectable intraocular lens materials based upon hydrogels. Biomacromolecules, 2001, 2 (3): 628-634.

[31] Guo J, Liu X, Jiang N, et al. Highly stretchable, strain sensing hydrogel optical fibers. Advanced Materials, 2016, 28 (46): 10244-10249.

[32] Choi M, Humar M, Kim S, et al. Step-index optical fiber made of biocompatible hydrogels. Advanced Materials, 2015, 27 (27): 4081-4086.

[33] Yuk H, Lin S, Ma C, et al. Hydraulic hydrogel actuators and robots optically and sonically camouflaged in water. Nature communications, 2017, 8 (1): 14230.

[34] Blaiszik B J, Kramer S L, Olugebefola S C, et al. Self-healing polymers and composites. Annual review of materials research, 2010, 40: 179-211.

[35] Dai X, Zhang Y, Gao L, et al. A Mechanically Strong, Highly Stable, Thermoplastic, and self-healable supramolecular polymer hydrogel. Advanced Materials, 2015, 27 (23): 3566-3571.

[36] Jia H, Huang Z, Fei Z, et al. Unconventional tough double-network hydrogels with rapid mechanical recovery, self-healing, and self-gluing properties. ACS applied materials & interfaces, 2016, 8 (45): 31339-31347.

[37] Shao C, Chang H, Wang M, et al. High-strength, tough, and self-healing nanocomposite physical hydrogels based on the synergistic effects of dynamic hydrogen bond and dual coordination bonds. ACS applied materials & interfaces, 2017, 9 (34): 28305-28318.

[38] Han L, Yan L, Wang K, et al. Tough self-healable and tissue-adhesive hydrogel with tunable multifunctionality. NPG Asia Materials, 2017, 9 (4): e372.

[39] Zhang H, Wang C, Zhu G, et al. Self-healing of bulk polyelectrolyte complex material as a function of pH and salt. ACS Applied Materials & Interfaces, 2016, 8 (39): 26258-26265.

[40] Darnell M C, Sun J Y, Mehta M, et al. Performance and biocompatibility of extremely tough alginate/polyacrylamide hydrogels. Biomaterials, 2013, 34 (33): 8042-8048.

[41] Xie Z, et al. Self-healable, super tough graphene oxide-poly (acrylic acid) nanocomposite hydrogels facilitated by dual cross-linking effects through dynamic ionic interactions. J Mater Chem B, 2015, 3 (19): 4001-4008.

[42] Kurt B, Gulyuz U, Demir D D, et al. High-strength semi-crystalline hydrogels with self-healing and shape memory functions. European Polymer Journal, 2016, 81: 12-23.

[43] Nakahata M, Takashima Y, Harada A. Highly flexible, tough, and self-healing supramolecular polymeric materials using host–guest interaction. Macromolecular rapid communications, 2016, 37 (1): 86-92.

[44] Neal J A, Mozhdehi D, Guan Z. Enhancing mechanical performance of a covalent self-healing material by sacrificial noncovalent bonds. Journal of the American Chemical Society, 2015, 137 (14): 4846-4850.

[45] Liu X., Yuk H, Lin S, et al. 3D printing of living responsive materials and devices. Advanced Materials, 2018, 30 (4): 1704821.

[46] Yamaguchi M, Ono S, Okamoto K. Interdiffusion of dangling chains in weak gel and its application to self-repairing material. Materials Science and Engineering: B, 2009, 162 (3): 189-194.

[47] Sirajuddin N, Jamil M. Self-healing of poly (2-hydroxyethyl methacrylate) hydrogel through molecular diffusion. Sains Malaysiana, 2015, 44 (6): 811-818.

[48] Zhang H, Xia H, Zhao Y. Poly (vinyl alcohol) hydrogel can autonomously self-heal. ACS Macro Letters, 2012, 1 (11): 1233-1236.

[49] Matsuda T, Kawakami R, Namba R, et al. Mechanoresponsive self-growing hydrogels inspired by muscle training. Science, 2019, 363 (6426): 504-508.

[50] Sun T L, Kurokawa T, Kuroda S, et al. Physical hydrogels composed of polyampholytes demonstrate high toughness and viscoelasticity. Nature materials, 2013, 12 (10): 932-937.

[51] Correa S, Grosskopf A K, Lopez Hernandez H, et al. Translational applications of hydrogels. Chemical Reviews, 2021, 121 (18): 11385-11457.

[52] Mandal A, Clegg J R, Anselmo A C, et al. Hydrogels in the clinic. Bioengineering & Translational Medicine, 2020, 5 (2): e10158.

[53] Chen M H, Chung J J, Mealy J E, et al. Injectable supramolecular hydrogel/microgel composites for therapeutic delivery. Macromolecular bioscience, 2019, 19 (1): 1800248.

[54] Mann J L, Anthony C Y, Agmon G, et al. Supramolecular polymeric biomaterials. Biomaterials science, 2018, 6 (1): 10-37.

[55] Webber M J, Appel E A, Meijer E, et al. Supramolecular biomaterials. Nature materials, 2016, 15 (1): 13-26.

第 8 章
水凝胶的动态模拟

8.1 光图案化和光化降解法
8.2 动态光度图形法
8.3 细胞响应反馈系统法
8.4 刺激响应——形态变形法
8.5 细胞介导牵引力引起的形态变形法
参考文献

无论是天然的还是人工合成的，没有任何一个现有的水凝胶系统是纯静态的，也就是说，具有动态演化特性的水凝胶系统普遍存在，大量的研究聚焦于以时间为变量的方式调节水凝胶行为的能力。在设计相关的系统时，研究人员强烈希望水凝胶微环境中会随时间而改变的内在嵌入部分（结构/成分）可以响应外部刺激，不论该刺激是以明确方式外部施加的还是由水凝胶负载物自身施加的。在过去十年中，各种将动态的外部信号结合到水凝胶平台中以精确调节负载物-水凝胶基质之间的相互作用的方法被开发出来[1,2]，为构建软执行器和柔性机器人技术等提供了新思路[3-6]。

8.1 光图案化和光化降解法

　　光图案化是一种用于水凝胶动态调控的成熟技术，该方法通过生物活性分子的受控空间固定化特性，在水凝胶中引入异质成分[7-9]。只需要通过光刻将特定的生化物质添加到水凝胶基质中，使水凝胶局部包含信号分子，可以让细胞在特定条件下做出反应。这些信号分子可以是生长因子、蛋白质、激素、细胞黏附肽、细胞排斥部分等[9]。值得注意的是，该行为的逆过程（光化降解法）——从水凝胶基质中按需释放某些成分——有时也需要触发水凝胶随时间变化而出现的行为变化。在实现该能力的过程中，光降解（与光交联相反）是指在包含双光子照射下进行光解的光不稳定成分（结构）的基础上发展起来的（图8.1）。水凝胶的这些成分（结构）若具有生物相容性，就能在生物活性分子或细胞存在的情况下实现水凝胶的时间依赖性光降解。例如，Kloxin等[10]通过具有细胞相容性的大分子单体的快速聚合来原位远程操纵水凝胶的特性，从而构建了可光降解的聚乙二

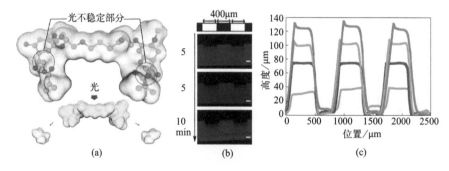

图8.1　（a）光不稳定部分进行光降解[11]；（b）水凝胶在照射后出现的表面侵蚀（比例尺：100μm）；（c）水凝胶通过蒙面泛光照射在空间上受到侵蚀，其中特征尺寸在不同照射时间后使用轮廓测量法进行量化[10]

醇基水凝胶体系。水凝胶相应特性的控制被证明可以往水凝胶中引入时间变化、形成任意形状特征、按需释放悬垂结构的功能。在包含封装细胞的水凝胶内的光降解通道允许细胞在其中迁移，生化凝胶成分的时间依赖特性可以被用来影响封装干细胞的软骨分化，这为实时操纵可光降解凝胶的材料结构特性或化学成分提供了动态环境，使得该水凝胶可以响应有关活细胞的外在刺激，并实现从药物输送载体设计到组织工程系统的应用。

8.2 动态光度图形法

将光图案化和光降解法相结合，可以实现对水凝胶系统中空间图案动力学的完全控制[12,13]。光可逆图案化策略是通过光介导的连接来确定生物分子在水凝胶中的空间分布，暴露于不同的光中进行水凝胶中相应生物分子的释放（图8.2）。DeForest等[13]利用两种生物正交光化学法实现了生物活性全长蛋白质的可逆固定，并在充满细胞的合成仿生水凝胶支架内实现了空间和时间的控制。所形成的光脱保护-肽连接序列，可以控制一定数量的蛋白质锚定在水凝胶三维矩阵的不同子结构内，而邻硝基苄酯光裂解反应有助于蛋白质从水凝胶支架上脱除。通过使用这种方法来模拟细胞外基质蛋白和玻连蛋白的状态，DeForest等以空间调控的方式实现了人的间充质干细胞向成骨细胞的可逆分化，未来可以用于探测和指导细胞生理学响

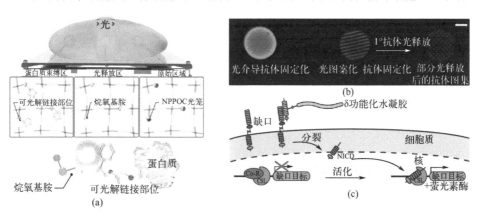

图8.2 （a）动态光度图形法[11]。（b）图案化的单抗用荧光双抗可视化，随后被光释放以形成二级图案，比例尺：3mm。（c）Tethered Delta 与表面缺口受体相互作用，触发两个蛋白水解裂解，从质膜释放缺口胞内结构域（NICD）。NICD 经历核易位并将 CSL 转录抑制因子转化为激活因子。这种激活促进了缺口介导的信号传导，并且在使用的 U2OS 缺口报告细胞系的情况下，还促进了萤光素酶的表达[13]

应动态生化信号的变化。此外，基质力学以及可用于调整封装细胞的行为等水凝胶内部生物物理线索的动态调控机制的研究也受到广泛关注[14,15]。

8.3 细胞响应反馈系统法

细胞可能是另一种外部刺激源，可触发它们所在的水凝胶基质发生动态变化[16-18]。在人体中，细胞外基质（ECM）的一个关键特征在于它对由驻留细胞引发的蛋白水解的敏感性，以促进它们的移动和随后的微环境重塑。为了在水凝胶中实现这种功能，通常可以将特殊设计的对酶敏感的分子或结构共聚到水凝胶中，以便允许局部蛋白酶介导的高分子网络实现降解[17]。结合生长因子的螯合成分，这些受细胞控制的水凝胶基质能够按需可逆地固定生长因子[19]。最近，基于蛋白酶介导的水凝胶降解概念又向前迈进了一步，研究人员通过水凝胶基质中的酶不稳定成分以及共同嵌入水凝胶基质中的酶抑制剂的联合作用实现了负反馈回路[18]。其中，酶敏感肽［例如基质金属蛋白酶（MMP）］作为水凝胶网络的连接体，而酶抑制剂［例如 MMP 的组织抑制剂（TIMP）］也会受到酶敏感肽因环境刺激解离的作用而释放，因此也可破坏相应的水凝胶网络。这种包含酶-抑制剂的水凝胶体系可以实现对水凝胶体系局部激活和酶活性抑制的诱导平衡，从而调节水凝胶的降解（图8.3）。与传统的无限制刺激反应相比，反馈系统的引入在实现更精确的水凝胶基质特性的操作方面展现出更大的优势。这种基于细胞反应性MMP的机制适用于PEG-肝素等其他类型水凝胶的构建[20]。另外，在水凝胶中不引入刺激响应成分的情况下，通过在水凝胶形成之前对大分子进行选择性修饰，制备可以具有响应能力的水凝胶。当肝素被选择性脱硫时，与原始肝素相比，所得水凝胶显示亲和固定生长因子的释放曲线会发生变化[21]。该策略可以扩展到水凝胶的化学修饰，以便进一步实现更好的生物分子螯合和释放的动态控制。

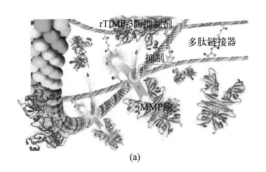
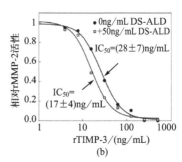

(a)　　　　　　　　(b)

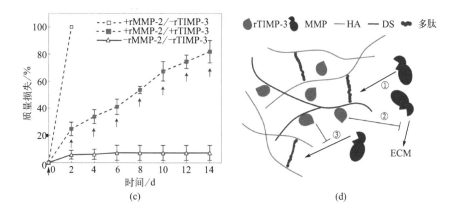

图 8.3 （a）细胞响应反馈系统法[11]。（b）通过其抑制重组 MMP-2（rMMP-2）溶液的能力来测量重组 TIMP-3（rTIMP-3）活性。（c）具有（实心符号）和没有（空心符号）封装的 rTIMP-3 的水凝胶与具有（正方形）或没有（三角形）rMMP-2 一起孵育，其中封装的 rTIMP-3 减弱了 rMMP-2 介导的水凝胶降解。（d）MMP 降解水凝胶交联①，释放多糖结合的 rTIMP-3，从而抑制局部 MMP 的活性②，并进一步减弱水凝胶的降解③ [18]

8.4 刺激响应——形态变形法

形态的变化是生物体的一个重要特征。对于自然界的生物而言，随着有机体的发育，它会沿着特定的路径改变形态，以产生最终的目标形状和结构，并以协调的方式发挥作用，从而使整个有机体发挥功能。因此，我们可以通过选取构成水凝胶的特定高分子，并对其修饰改性，以确保在其在受到湿度[4,22]、温度[23]、pH[24,25]、离子强度[25,26]、磁性[27,28]、光[5]等外界刺激后具有仿生形状变形能力[6]。仿生变形水凝胶的构建，通常需要包含具有不同响应度的异质材料或多层配置才能实现相应的动态形状变化。例如，由两种具有独特溶胀能力的聚合物（如 PEG 和海藻酸盐）形成的双层水凝胶在吸水时可以呈现定向弯曲[22]，热敏和热惰性水凝胶组成的材料在温度高于或低于热敏水凝胶组分成型的温度下可以调节水凝胶结构的双向弯曲（图 8.4）[23]。此外，将 pH 敏感部分嵌入水凝胶中可以实现 pH 介导的溶胀或消溶胀[24]。尽管这些模型依赖于构成水凝胶的高分子的固有特性，但通过将磁性纳米粒子[28]、碳纳米材料[5]等功能材料添加到水凝胶基质中，可以实现分别由磁场和光引起的局部加热从而实现相应的驱动。

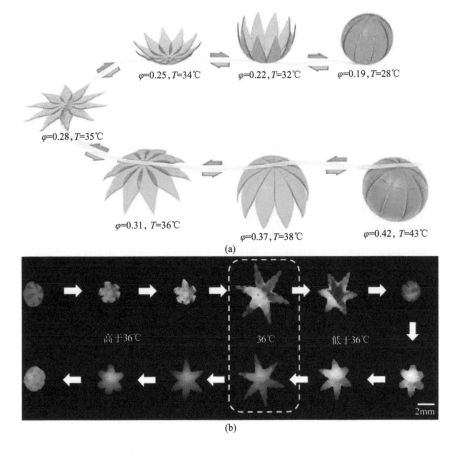

图8.4 （a）温度响应法[11]。（b）当温度降至36℃以下时，在外部开口上带有聚富马酸丙烯酯（PPF）片段的闭合夹持器的系列图像，然后自行折叠成闭合夹持器，但带有PPF片段朝向内侧[23]

8.5 细胞介导牵引力引起的形态变形法

　　细胞介导的牵引力有助于水凝胶的形状变形[29]。例如，精心设计的植入细胞的蛋白微结构能够以折纸的方式自行折叠成预定义的结构[30]，或自发收缩的细胞能够使其所在的水凝胶基质发生机械运动而形成软生物致动器（图8.5）[31,32]。早期的生物致动器主要是基于弹性体材料，而最近水凝胶材料因其良好的生物相容性而受到越来越多的关注。通常采用肌细胞（骨骼肌细胞或心肌细胞）以及能够在这些细胞之间进行信号传输的水凝胶的组合，构筑生物致动器。碳纳米管复合明胶甲基丙烯酰水凝胶被加工成厘米级的柔性基底，附着于心肌细胞表面并实现同步跳动，使得其表现出有节奏的收缩和伸展行为[33,34]。工程化的骨骼肌，当与生物打印的水凝胶框架结合时，也可以在人为设定的条件下驱动[35,36]。这些依赖于细胞牵引力的生物致动器可以通过电信号[33,34,37]、光信号[32,35]等外部刺激进行远程控制。

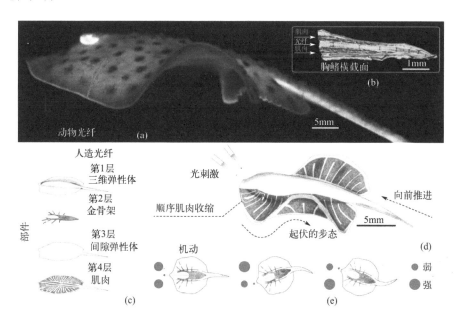

图 8.5　小白鲼的（a）游泳时的照片，（b）肌肉骨骼结构，（c）四层身体结构示意图，（d）刺激作用-响应概念示意图，（e）趋光控制[32]

水凝胶的结构是动态可调节的，水凝胶材料的设计需结合时间维度变量的影响因子。例如，当我们采用 3D 打印技术构建相应功能水凝胶的时候，若考虑到时间参数的影响，可选择 4D 打印。

参考文献

[1] Burdick J A，Murphy W L. Moving from static to dynamic complexity in hydrogel design. Nature communications，2012，3（1）：1269.

[2] Tibbitt M W，Anseth K S. Dynamic microenvironments：the fourth dimension. Science translational medicine，2012，4（160）：160ps24.

[3] Kim J，Hanna J A，Byun M，et al. Designing responsive buckled surfaces by halftone gel lithography. Science，2012，335（6073）：1201-1205.

[4] Jeong K U，Jang J H，Kim D Y，et al. Three-dimensional actuators transformed from the programmed two-dimensional structures via bending，twisting and folding mechanisms. Journal of Materials

Chemistry, 2011, 21 (19): 6824-6830.

[5] Wang E, Desai M S, Lee S W. Light-controlled graphene-elastin composite hydrogel actuators. Nano letters, 2013, 13 (6): 2826-2830.

[6] Ionov L. Hydrogel-based actuators: possibilities and limitations. Materials Today, 2014, 17 (10): 494-503.

[7] Loessner D, Meinert C, Kaemmerer E, et al. Functionalization, preparation and use of cell-laden gelatin methacryloyl-based hydrogels as modular tissue culture platforms. Nature protocols, 2016, 11 (4): 727-746.

[8] Schuurman W, Levett P A, Pot M W, et al. Gelatin-methacrylamide hydrogels as potential biomaterials for fabrication of tissue-engineered cartilage constructs. Macromolecular bioscience, 2013, 13 (5): 551-561.

[9] Wylie R G, Ahsan S, Aizawa Y, et al. Spatially controlled simultaneous patterning of multiple growth factors in three-dimensional hydrogels. Nature materials, 2011, 10 (10): 799-806.

[10] Kloxin A M, Kasko A M, Salinas C N, et al. Photodegradable hydrogels for dynamic tuning of physical and chemical properties. Science, 2009, 324 (5923): 59-63.

[11] Zhang Y S, Khademhosseini A. Advances in engineering hydrogels. Science, 2017, 356 (6337): eaaf3627.

[12] DeForest C A, Anseth K S. Cytocompatible click-based hydrogels with dynamically tunable properties through orthogonal photoconjugation and photocleavage reactions. Nature chemistry, 2011, 3 (12): 925-931.

[13] DeForest C A, Tirrell D A. A photoreversible protein-patterning approach for guiding stem cell fate in three-dimensional gels. Nature materials, 2015, 14 (5): 523-531.

[14] Khetan S, Guvendiren M, Legant W R, et al. Degradation-mediated cellular traction directs stem cell fate in covalently crosslinked three-dimensional hydrogels. Nature materials, 2013, 12 (5): 458-465.

[15] Yang C, Tibbitt M W, Basta L, et al. Mechanical memory and dosing influence stem cell fate. Nature materials, 2014, 13 (6): 645-652.

[16] Sakiyama-Elbert S E, Panitch A, Hubbell J A. Development of growth factor fusion proteins for cell-triggered drug delivery. The FASEB Journal, 2001, 15 (7): 1300-1302.

[17] Lutolf M P, Lauer-Fields J L, Schmoekel H G, et al. Synthetic matrix metalloproteinase-sensitive hydrogels for the conduction of tissue regeneration: engineering cell-invasion characteristics. Proceedings of the National Academy of Sciences, 2003, 100 (9): 5413-5418.

[18] Purcell B P, Lobb D, Charati M B, et al. Injectable and bioresponsive hydrogels for on-demand matrix metalloproteinase inhibition. Nature materials, 2014, 13 (6): 653-661.

[19] Watarai A, Schirmer L, Thönes S, et al. TGFβ functionalized starPEG-heparin hydrogels modulate human dermal fibroblast growth and differentiation. Acta biomaterialia, 2015, 25: 65-75.

[20] Taubenberger A V, Bray L J, Haller B, et al. 3D extracellular matrix interactions modulate tumour cell growth, invasion and angiogenesis in engineered tumour microenvironments. Acta biomaterialia, 2016, 36: 73-85.

[21] Zieris A, Dockhorn R, Röhrich A, et al. Biohybrid networks of selectively desulfated glycosaminoglycans for tunable growth factor delivery. Biomacromolecules, 2014, 15 (12): 4439-4446.

[22] Zhang L, Liang H, Jacob J, et al. Photogated humidity-driven motility. Nature communications, 2015, 6 (1): 7429.

[23] Breger J C, Yoon C, Xiao R, et al. Self-folding thermo-magnetically responsive soft microgrippers. ACS applied materials & interfaces, 2015, 7 (5): 3398-3405.

[24] Yu Q, Bauer J M, Moore J S, et al. Responsive biomimetic hydrogel valve for microfluidics. Applied Physics Letters, 2001, 78 (17): 2589-2591.

[25] Ma C, Li T, Zhao Q, et al. Supramolecular Lego Assembly Towards Three-Dimensional Multi-Responsive Hydrogels. Advanced Materials, 2014, 26 (32): 5665-5669.

[26] Wu Z L, Moshe M, Greener J, et al. Three-dimensional shape transformations of hydrogel sheets induced by small-scale modulation of internal stresses. Nature communications, 2013, 4 (1): 1586.

[27] Kokkinis D, Schaffner M, Studart A R. Multimaterial magnetically assisted 3D printing of composite materials. Nature communications, 2015, 6 (1): 8643.

[28] Zhao X, Kim J, Cezar C A, et al. Active scaffolds for on-demand drug and cell delivery. Proceedings of the National Academy of Sciences, 2011, 108 (1): 67-72.

[29] Kim S, Laschi C, Trimmer B. Soft robotics: a bioinspired evolution in robotics. Trends in biotechnology, 2013, 31 (5): 287-294.

[30] Kuribayashi-Shigetomi K, Onoe H, Takeuchi S. Cell origami: self-folding of three-dimensional cell-laden microstructures driven by cell traction force. PloS one, 2012, 7 (12): e51085.

[31] Nawroth J C, Lee H, Feinberg A W, et al. A tissue-engineered jellyfish with biomimetic propulsion. Nature biotechnology, 2012, 30 (8): 792-797.

[32] Park S J, Gazzola M, Park K S, et al. Phototactic guidance of a tissue-engineered soft-robotic ray. Science, 2016, 353 (6295): 158-162.

[33] Shin S R, Jung S M, Zalabany M, et al. Carbon-nanotube-embedded hydrogel sheets for engineering cardiac constructs and bioactuators. ACS nano, 2013, 7 (3): 2369-2380.

[34] Shin S R, Shin C, Memic A, et al. Aligned carbon nanotube-based flexible gel substrates for engineering biohybrid tissue actuators. Advanced functional materials, 2015, 25 (28): 4486-4495.

[35] Raman R, Cvetkovic C, Uzel S G, et al. Optogenetic skeletal muscle-powered adaptive biological machines. Proceedings of the National Academy of Sciences, 2016, 113 (13): 3497-3502.

[36] Cvetkovic C, Raman R, Chan V, et al. Three-dimensionally printed biological machines powered by skeletal muscle. Proceedings of the National Academy of Sciences, 2014, 111 (28): 10125-10130.

[37] Shin S R, Zihlmann C, Akbari M, et al. Reduced graphene oxide-gelMA hybrid hydrogels as scaffolds for cardiac tissue engineering. Small, 2016, 12 (27): 3677-3689.

第9章

仿生智能水凝胶软执行器及其应用

9.1 自然界中的刺激-响应驱动行为
9.2 人造刺激-响应性水凝胶执行器
9.3 仿生智能水凝胶执行器的应用
参考文献

大自然赋予无数生物体以刺激-响应的驱动行为来实现其基本生理活动的能力。受这些有趣的生物结构的启发，人们开发了各种具有可控性、快速响应性以及韧性的仿生水凝胶致动器。在本章中，我们将介绍自然界中一些典型的刺激-响应行为的内在策略以及具有相关特性的仿生水凝胶致动器等，目的是突出最近的进展，找到共同点，并讨论根本差异以确定该领域当前的挑战和未来的方向。

9.1 自然界中的刺激-响应驱动行为

大自然中存在许多不同的生物，如章鱼[1-3]、水母[4,5]、捕蝇草[6-8]和含羞草[9-11]等，它们能够通过对刺激的响应实现运动、捕食、繁殖等活动。基于这些有趣的现象，人们进行了许多尝试，开发出了人工响应性水凝胶执行器，包括人工肌肉[12-14]、智能执行器[15-18]和软体机器人[19-22]等。水凝胶是一类高度交联的亲水性网络和高含水量的材料，通过对其化学、结构和功能进行改性，可以开发出对刺激响应的人工软性执行器。然而，大多数的水凝胶因其网络和交联的不均匀分布而非常脆弱，并且由于缺乏能量耗散系统，在受到长期机械负荷时会变得越来越脆，这限制了仿生水凝胶执行器的发展和使用。长期以来，设计增强和能量耗散结构，以增加水凝胶的强度、韧性和抗疲劳性，一直是构建可以实现类似于自然生物的致动行为和能力水凝胶所关注的重点。此外，通过合理的响应性设计，水凝胶执行器也可以实现复杂的形状变化和运动过程。具有耐久性和响应性的水凝胶，由于其化学和结构的多样性，不仅表现出优异的柔韧性和生物相容性，还能够通过快速响应外部刺激来适应复杂的环境[23-25]。

从材料科学研究和实际应用的角度来看，传统的刺激响应性水凝胶执行器由简单高分子组成，存在诸多不足，如力学性能弱（强度＜50kPa，刚度＜10kPa，韧性＜10J/m^2），反应速度慢（如传统的 PNIPAM 水凝胶通常需要几分钟到几小时）以及环境适应性差，因此严重限制了它们在软执行器方面的实际应用。相比之下，具有复杂多组分和多尺度分层结构的生物组织的天然水凝胶具有出色的力学性能。例如，松果具有高的韧性模量和强度值，分别为 1.21GPa 和 25.24MPa。受这些软组织极端力学性能的启发，已经研发出许多具有出色力学性能的坚韧水凝胶执行器，为指导设计和加工具有机械稳健性的坚韧水凝胶奠定了坚实的基础，进而扩大了它们在软执行器中的功能应用[25-27]。通过进一步的化学设计或改变水凝胶网络结构，可以使水凝胶在暴露于温度[28-30]、pH[31,32]、光[33-36]、磁场[37,38]、电场[39,40]和氧化还原[41-43]等外界刺激时具有响应能力。基于响应性水凝胶驱动行为的柔性智能设备研究近年来受到越来越多的关注[44-46]。然而，如何有效地将响应性和增韧机制结合起来，开发能够承受重复负载的坚韧响应性水凝胶，在实际应用中仍然是一个关键的难点。典型的构建方法包括在同一设备上集成具有不同性质的水凝胶，在单个水凝胶中构建精确可控的性质分布，以及在受到外部刺激时实现编程变形或运动[47-49]。

本节将介绍仿生水凝胶执行器件，这些器件在受到特定刺激后表现出刺激响应性驱动行为，以及不同的刺激手段带来的优势和可能的应用场景（图9.1）。

图9.1　响应性水凝胶执行器的现有研究：从天然到仿生人工智能响应水凝胶及可能的应用范围

9.1.1　基于细胞渗透压变化实现的驱动行为

许多植物在受到周围环境的刺激后，可以通过细胞内离子的迁移或膜渗透性的变化来触发和控制启动。例如，维纳斯捕蝇草可以瞬间闭合叶子来捕捉昆虫[图9.2（a）]。当接受刺激时，维纳斯捕蝇草的叶子会与环境交换水分（吸收/释放水分），引起渗透压的变化，从而产生不同的膨胀压力来控制叶子的开放和闭合状态。捕蝇草闭合是由同时发生的外表皮扩张和内表皮收缩引起的，该运动不受中叶的主动影响。此外，由于各层之间的内部水压差（或称预应力），捕蝇草处于随时可以闭合的状态，这对于实现快速闭合至关重要。捕蝇草圈闭合的行为（当预应力部分释放时），可以表明水从正面到背面的运动可能导致捕集器变得更窄。间隙水通过弹性组织的运动在内部抑制了捕集器的快速关闭动作。叶子的大幅弯曲很可能是由残留的水位移造成的。诱使捕蝇草的两块不同的叶瓣相互闭合，一旦发生这种变化，双曲叶的几何形状提供了弹性能量储存和释放的机制，叶子的水合性质诱发了快速阻尼，这对有效捕捉猎物同样至关重要[50-53]。类似地，含羞草也可以在被触摸后关闭叶子［图9.2（b）］。然而，与捕蝇草依赖环境水交换不同，含羞草叶片的运动是通过细胞内水分的重新分配引起的可逆渗透压变化实现的[54,55]。

图 9.2 （a）食虫植物捕蝇草处于开启和关闭状态[50]；（b）含羞草叶在触摸前后的状态[54]

9.1.2 基于纤维素原纤维结构不均匀膨胀实现的驱动行为

除了渗透压的变化，细胞壁中纤维素纤维结构的不均匀膨胀也可以引发驱动行为。松果［图9.3（a）］，由两种不同方向的纤维素纤维组成；在潮湿的环境中，纵向纤维素纤维的长度增加，而横向纤维则增加宽度，导致体积不均匀膨胀，最终导致其开放[56-60]。同样，小麦芒有不同细胞壁结构的组织类型，其内部称为"帽"，朝向另一麦芒；其外部称为"脊"，朝向外面。在"帽"部，细胞壁具有几乎平行于细胞轴的定向纤维素纤维，使得芒的轴对湿度变化不太敏感。然而，在"脊"的下部，纤维素纤维是随机分布的，湿度的变化可以引起各向同性的收缩或膨胀，最终导致它的开合［图9.3（b）］[61-63]。

图 9.3 （a）松果在湿润和干燥状态下的形态[60]；（b）小麦芒在潮湿和干燥条件下的形态[63]

9.1.3 基于可逆弱键的断裂/生成实现的驱动行为

很多棘皮动物依靠它们皮肤组织中的纳米复合材料来加强高纵横比胶原纤维

的黏弹性基质,实现快速可逆的硬度变化。例如,海参可以硬化以防御外部威胁(图9.4)[64-66],其皮肤细胞释放的可溶性大分子可以促进皮肤组织中纤维相互作用的可逆调节[65]。这主要通过与相邻增强纤维表面的蛋白聚糖形成临时的弱键来实现,显著增加了皮肤组织的弹性模量[65,67,68]。

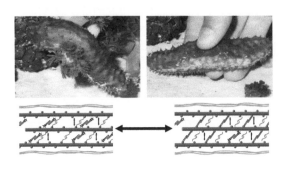

图 9.4 海参处于松弛和硬化状态[66]

9.1.4 基于微观结构变化实现的驱动行为

刺激响应的结构变化、结构单元的间距或平均折射率,使一些生物(如变色龙)具有动态、快速和可逆的结构色转变行为。变色龙可以调节其皮肤色调,以适应不断变化的环境。这个过程在几分钟内发生且是可逆的[69-73]。如图9.5(a)所

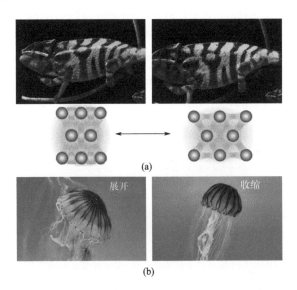

图 9.5 (a)成年变色龙在从静止状态到兴奋状态的颜色变化[74];
(b)水母的膨胀和收缩运动[75]

示,在防御(兴奋)状态下,成年雄性变色龙会将其皮肤的背景颜色从绿色变成黄色。这主要是因为两层虹彩皮细胞中含有鸟嘌呤晶体。上层的虹彩细胞由密集排列的直径为 127.4μm 的鸟嘌呤晶体组成。这些晶体排列成三角晶格。鸟嘌呤晶体(折射率为 1.83)与细胞质(折射率为 1.33)结合形成光子晶体结构。鸟嘌呤晶体的平均间距在静止状态(绿色皮肤)下比兴奋状态(黄色皮肤)下小 30%,而表面虹彩细胞中的晶体大小没有变化。当外部环境刺激变色龙时,其真皮被拉伸至兴奋状态。此时,纳米晶体的晶格间距增大,导致偏好反射光从短波长向长波长移动,颜色变化相应发生 [72-74]。水母通过其独特的体结构连续地"钟形"收缩,实现推进和捕食 [图 9.5 (b)] [75-78]。不断的"钟形"收缩可以帮助水母将肌肉周期性收缩产生的能量传递给周围的液体以产生推进力。此外,肌肉的高动作电位也可以使游泳肌肉处于兴奋状态并长时间收缩,令弹性结构在该状态下完全变形并储存大量应变能 [79,80]。

9.1.5 基于软结构的收缩/拉伸实现的驱动行为

自然界中,许多动物利用其软组织的可变形性在复杂环境中实现有效运动。章鱼的触臂具有出色的可控性,可伸展至原始长度的两倍,方便触及狭小裂缝中的猎物(图 9.6)。此外,章鱼还可以激活特定肌肉产生局部硬度,从而对环境施力。由于章鱼完全是软体且没有内外骨骼,它的触臂具有高度的柔韧性;该特性基于其独特的肌肉排列方式,由三种类型的肌肉结构组成:沿着触臂长度轴向延伸的纵向肌肉,斜向环绕触臂的倾斜肌肉和垂直于触臂轴线的辐射肌肉 [73,81]。沿着中央结缔组织分布的纵向和辐射肌肉提供了触臂伸长和缩短的对抗性收缩力,而倾斜肌肉则产生扭转力。这三种肌肉的协同作用使得诸如伸长、缩短和弯曲等动作成为可能,而这些基本动作的进一步组合则产生了更多种复杂的活动 [82-84]。

图 9.6 章鱼伸展触手抓取物体的状态 [84]

9.2 人造刺激-响应性水凝胶执行器

许多自然系统能够不断响应和适应周围环境的变化。生物体可以通过对刺激的感知（日常光照、温度变化以及周围的生化信号等），并产生对应的反应来调整它们的形状或位置以躲避天敌或实现更好的营养获取 [75,85]。这些响应和变形行为的内在原理为柔性电子、软体机器人、生物医学以及其他软物质技术等带来了先进的设计思路和指导思想。因此，越来越多的刺激响应材料被广泛用于开发具有变形和运动能力的人工智能控制系统。其中，水凝胶因其与软生物组织的生物学相似性、对多种刺激的响应性以及优异的塑形性，在相关领域具有广阔的应用前景。一般来说，具有刺激响应功能的水凝胶执行器的内在结构在受到外部异质刺激（包括温度、光、酸碱度、电场等）时可以表现出显著的体积/相变行为以及可逆的膨胀/收缩特性，这就是水凝胶执行器可以实现运动的关键作用机制。在此，我们将详细地介绍不同异质刺激诱导下的水凝胶执行器响应行为，包括温度、光、磁、电、酸碱度、离子、湿度、溶剂等体系（图9.7），并讨论这些不同控制系统的优缺点。

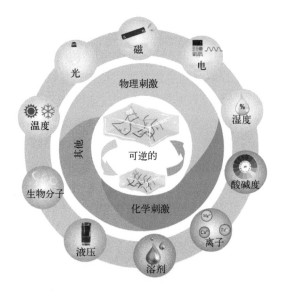

图 9.7　具有刺激响应性能的水凝胶执行器示意图

刺激响应执行器可以在各种刺激（如温度、光、磁、电、湿度、酸碱度、离子、溶剂和其他刺激）作用下展现出明显的体积/相变，并呈现可逆的膨胀/收缩行为 [86]

9.2.1　热响应

热刺激-响应体系是人造智能系统广泛使用和研究的内容之一，其在水凝胶执行器的设计和制备中展现出了不可替代的作用。以水凝胶为代表的热响应性软材料可以响应环境温度的变化，并通过临界相变温度下的溶胶-凝胶相/体积变化行为来实现相应的驱动（动作）[87,88]。从本质上讲，这种溶剂的吸收和排出的行为是由水凝胶内部焓和熵的平衡状态，基于温度的分子重组，以及高分子链与溶剂之间的相互作用共同决定的，从而在宏观上表现出基体体积的膨胀或收缩的行为[89-91]。通常，热响应水凝胶在 LCST 下表现出负热敏感性，然而当温度升至 UCST 时则会转变为正热敏感性[92]，这两种不同的响应特征进一步拓展了水凝胶执行器的类型和应用场景。

9.2.1.1 LCST 响应水凝胶执行器

响应性高分子的 LCST 是其下临界溶解温度，低于该温度时所有的高分子组分将会完全混溶。在众多的 LCST 响应性高分子中，聚 N-异丙基丙烯酰胺（PNIPAM）因其具有低毒性、电中性、化学稳定性、低 pH 依赖性、接近体温的 LCST（约 32℃）和大范围的可调溶胀比（范围为 2～20）等特性引起了广泛的关注。当环境温度从低于 LCST 升至高于 LCST 时，PNIPAM 水凝胶会从完全溶解在水中到逐渐收缩，直至最终从水溶液中析出（图9.8）；而当逐步降低环境温度直至低于 LCST 时，PNIPAM 会吸收回其周围的溶剂并膨胀至其溶于水的状态，表明其在热响应下的体积相变具有可逆性[93]。为了方便理解，我们将通过几个例子说明典型的 LCST 高分子（聚 N-异丙基丙烯酰胺水凝胶）执行器的一般物理机制，并讨论这种 LCST 响应水凝胶执行器的活动行为。

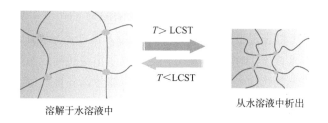

图 9.8　PNIPAM 水凝胶的致动机理示意图[86]

He 等[92]报道了一种具有微结构热响应性的聚 N-异丙基丙烯酰胺水凝胶执行器，它不仅能够实现温度控制的传感和驱动，还可以精确操控化学-机械-化学反馈回路的"开"与"关"（图9.9）。当温度低于 LCST 时，水凝胶执行器开始膨胀，带有催化剂功能化尖端的嵌入式微结构将会进入试剂层，并进一步引发放热反应。当化学反应放热引起的温度高于 LCST 时，热量会触发水凝胶收缩，使得微结构缩回从而失去催化作用以中断反应；当温度再次降至 LCST 以下时，自调节循环重新开始（图9.9）。在这种微结构水凝胶执行器中，基于水凝胶负载催化剂的微结构双层膜与含有反应物的"营养"层原本是分离的，利用响应温度变化的水凝胶的动态行为实现微结构进入或离开营养层的可逆驱动，从而精准操控化学反应的"开"与"关"，为实现高精度程序化控制的化学反应装置提供了思路。

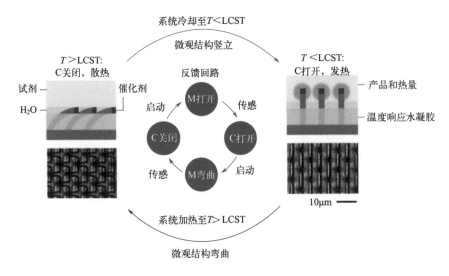

图 9.9　自调节机械化学适应性可重构性能的 LCST 调节水凝胶微结构[92]

Kim 等[93]开发了一种基于层状热响应 PNIPAM 结合单层钛酸盐纳米薄片（TINS）的智能水凝胶执行器，通过含 TINS 的 PNIPAM 收缩和膨胀，可以实现温度控制的各向异性变形（图 9.10）。与基于 PNIPAM 的传统均质智能水凝胶相似，该体系同样可以表现出均匀的加热诱导收缩和冷却诱导膨胀。基于 TINS 的水凝胶在平行和垂直于 TINS 平面的方向之间显示出逆热响应函数（即加热引起的膨胀和冷却引起的收缩）和差异膨胀／收缩（即各向异性）。随着温度的升高，PNIPAM 的静电介电常数增加，这是由于 TINS 之间的静电排斥力的增加而导致内部释放水分子的减少。这项工作展示出基于 PNIPAM 的 LCST 凝胶执行器更复杂和多样的形式，也为这类水凝胶提供了更多应用的可能。

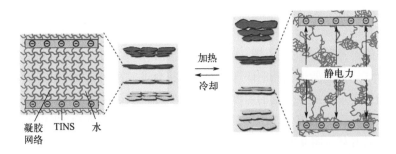

图 9.10　基于热响应变形的 PNIPAM/TINS 水凝胶执行器的示意图[93]

Nojoomi 等[94]使用数字光 4D 打印的方法来制备具有程序化运动形态 3D 结构 LCST 水凝胶执行器。将水凝胶前体暴露于数字光中，同时监测曝光时间（t_{ex}），通过数字光调节 4D 打印图案化 PNIPAM 网络的整体密度，从而获得不同交联密度及变形能力的水凝胶执行器。水凝胶网络的交联密度随着曝光时间的增加而增加，并且密度的增加又降低了宏观膨胀和收缩的程度和速率。通过这种有效调节，他们获得了预编程 2D 水凝胶盘并变形为各种复杂的 3D 形态（图 9.11 顶部）。在体积相变温度（T_c，约 32.5℃）下，该结构在体积相变时表现出膨胀和收缩状态之间的可逆形状转变。因为该方法编码了基于 PNIPAM 的智能水凝胶中的空间和时间相关的扩展/收缩和规定的指标，所以制备的 3D 结构可以在其缩水状态下显示出可编程的时间顺序运动（图 9.11 底部）和形态，可用于设计和制造软机器人、执行器和人造肌肉等软物质系统的设计。

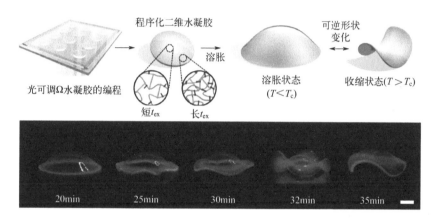

图 9.11 三维结构 PNIPAM 水凝胶执行器的示意图和照片（比例尺为 5cm）[94]

9.2.1.2 UCST 响应水凝胶执行器

不同于 LCST 水凝胶执行器，UCST 水凝胶执行器在温度升高时会膨胀。一般情况下，UCST 水凝胶执行器具有扩散网络特征，最常用的高分子基体主要是聚丙烯酰胺（PAM）和聚丙烯酸（PAAc）以及两者的复合物。当温度低于 UCST 时，水凝胶网络中的丙烯酸（AAc）和丙烯酰胺（AM）链之间表现为强相互作用，从而导致水凝胶体积收缩；当温度高于 UCST 时，AAc-AM 网络之间的氢键迅速解离，从而使得水凝胶的体积膨胀。此外，可以进一步在 UCST 共聚物分子链段中进行共聚或改性，以改变水凝胶的特性并实现想要的功能[86,95-97]。在接下来的内容中，我们将通过具体实例详细讨论 UCST 水凝胶执行器的作用机制。

Zhao 等通过不对称光聚合的方法制备了一种双层水凝胶执行器，包括一层具有 UCST 效应的 PAM/PAAc 层和一层没有响应功能的 PAM 层[95]。当温度低于 UCST 时，由于 PAM 和 PAAc 之间形成的大量氢键导致该层水凝胶开始收缩，而纯 PAM 层不受温度影响保持其原有状态，因而水凝胶执行器表现为自发变形（图 9.12）。该水凝胶执行器的驱动幅度可以进一步通过调节两层的组成比例来实现。这种不对称分布的高分子氢键概念可应用于其他 UCST 水凝胶的制备，并激发智能执行器的相关新发明。

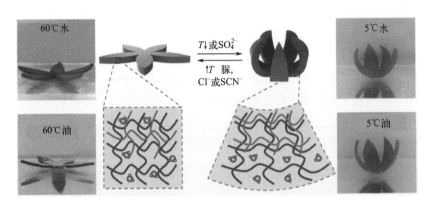

图 9.12　UCST 调控的水凝胶微结构的示意图，具有自调节的机械化学适应性重构性能[95]

此外，更多的尝试用于制备兼具 LCST 和 UCST 响应特性的智能水凝胶[98]。Li 等[99]利用聚 N- 丙烯酰胺（PNAGA）层和 PNIPAM 层合成了一种温度驱动的水凝胶执行器，这种执行器表现出 UCST 和 LCST 双重响应控制的特征，可以实现各种复杂的变形过程（图 9.13）。由于两种特定高分子在 UCST 和 LCST 的互补性温度响应行为，该水凝胶中的两层高分子在低温和高温下表现出完全相反的热响应膨胀和收缩特性，赋予水凝胶执行器快速的热响应和恢复能力。稳定和高效的响应特性使得这种水凝胶执行器可用于各种需要灵敏反应特性的操作，包括抓取、运输、释放物体或作为电路开关（例如打开和关闭发光二极管）等。

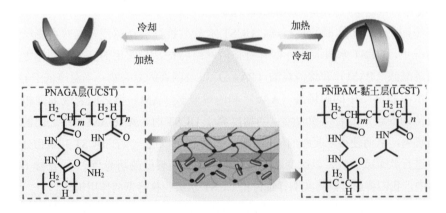

图 9.13　基于热响应变形的水凝胶执行器的示意图，具有 PNAGA 层和 PNIPAM- 黏土层[99]

在上述系统中,水凝胶的驱动是通过其恒定的体积收缩或膨胀实现的,具体表现为在响应温度下释放/吸收水,而这种行为就需要执行器浸泡在相应的溶液中,以实现灵活稳定的驱动行为,然而这一要求限制了水凝胶执行器更多的应用可能。为了解决这个问题,Zheng 等[100]报道了一种由 P(AAc-co-AM)层(UCST 水凝胶)和 PNIPAM 层(LCST 水凝胶)构成的双层水凝胶执行器,其灵感来自含羞草(图 9.14)。当环境温度低于 LCST 时,PNIPAM 水凝胶层保有大量的水,而 P(AAc-co-AM)层失去一部分水分,从而使 UCST 水凝胶侧出现弯曲变形。当温度同时高于 LCST 和 UCST 时,LCST 水凝胶失去水分,水被排放到 UCST 水凝胶层中,表现为体积塌陷,因此执行器沿 LCST 水凝胶方向弯曲。这种水凝胶执行器可以通过内部水分的传输实现露天环境下的驱动执行,从而摆脱了水凝胶执行器必须在相应的溶剂中实现驱动功能的限制,为其他更多应用场景水凝胶执行器的开发提供了思路。

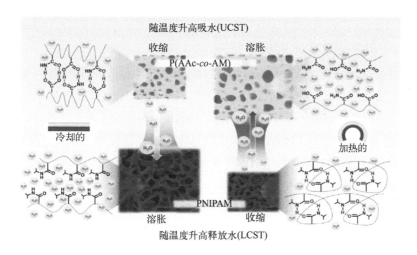

图 9.14 双层水凝胶执行器的示意图,包括 UCST 层和 LCST PNIPAM 层[100]

9.2.2 光响应

光刺激凭借其远距离、快速、无接触、灵敏等特点成为深受关注的刺激条件。一般而言,光响应水凝胶驱动器的主要驱动机制是可逆交联反应和光热激发,这两者都可以通过将光活性成分掺杂到水凝胶基质中来实现远程光激发行为(图 9.15)。在基于可逆交联反应机制的光响应水凝胶驱动器中,最常见的是光异构化和光循环两种方法。其中,水凝胶执行器的光异构化是指在光照射下诱导偶氮苯(顺/反式变化)和螺吡喃(开/闭环)等光引发剂实现水凝胶的驱动和变形。

而光循环控制过程是通过引入光不稳定的保护基团（如对羟基苯甲酰基和三苯基甲烷基团在受到光照时的交联和去交联行为）来实现水凝胶的可逆收缩和膨胀行为[96,100]。受益于光响应水凝胶执行器的发展，自愈水凝胶、遥控软机器人、分子机器和受控药物释放等典型的应用已经被开发出来。一般来说，由于波长和能量的差异，光刺激的来源可分为自然光、紫外光（UV）、近红外光（NIR）和红外光等多种形式。接下来我们将分别介绍自然光、紫外光和近红外光这三种光驱动的水凝胶执行器并详细讨论它们各自实现驱动的过程。

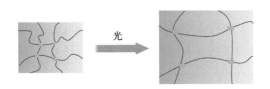

图 9.15　光敏感水凝胶致动机理的示意图[86]

9.2.2.1　自然光响应水凝胶执行器

自然光是一种最容易获得的刺激源，相比于其他光源，不需要任何复杂光源设备即可实现对水凝胶的驱动行为。基于此，有很多工作都集中在以自然光为刺激实现精确驱动的智能水凝胶。Xia 等[101]构建了一种掺杂了可见光响应的光热金纳米粒子（AuNPS）的 PNIPAM-AuNPS/PAM 薄膜水凝胶执行器。在光刺激下，这种水凝胶执行器内部的光热纳米粒子通过自身的光热效应产热，当温度超过 PNIPAM 层 LCST 后会使得水凝胶开始收缩，从而实现响应弯曲变形。如图 9.16 所示，将 PNIPAM-AuNPS/PAM 执行器固定在缓冲溶液中，当受到可见光照射时，水凝胶网络中的 AuNPS 可以吸收辐射能并转化为热能，从而提高水凝胶基体的温度以实现快速弯曲。当灯关闭时，执行器返回自然垂直状态。自然光的可逆驱动性能在远程精确控制过程显示出关键的价值。

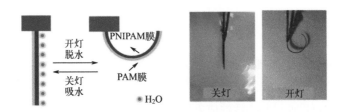

图 9.16　双水凝胶层的自然光调控示意图[101]

Yang 等[102]通过 3D 打印的方法成功地制备了以聚氨酯（PU）和炭黑（CB）为打印掺杂材料的自然光响应水凝胶执行器。受向日葵的启发，他们制备了光触发形状记忆复合人造多层向日葵，由黑色感光形状记忆材料打印出的向日葵花瓣、花蕊、花梗和聚乳酸打印出的花瓶组装而成（图9.17）。该人造向日葵的具体驱动过程与真正的向日葵运行机制相似。在光照刺激下，炭黑可以不断吸收太阳光能并转化为热能，当向日葵被加热到30℃（T_g）时，其花瓣由闭合状态转变为张开状态。该状态下的花瓣是柔软的并具有弹性，可以追踪光源的移动。此外，这种向日葵的变形是可逆的。例如，当向日葵的温度在没有外部光源的情况下逐渐冷却到T_g以下时，该人造花瓣会恢复到初始闭合状态并保持该形态不变，等有外界刺激时再次展现出变形行为。这种光敏仿生水凝胶执行器的开发为制备更复杂的光响应执行器提供了新思路。

图9.17 具有向日葵特性以响应自然光的水凝胶执行器的示意图（比例尺为2cm）[102]

9.2.2.2 紫外光响应水凝胶执行器

不同于自然光刺激，紫外光（UV）也被广泛应用于响应其驱动的分子和化合物中来实现有效的光驱动的可逆形状变化和位移。Sun等[103]报道了一种含有TINS和AuNPS的具有紫外线响应性的各向异性PNIPAM水凝胶执行器（图9.18），由AuNPS（光热剂），N,N'-亚甲基双丙烯酰胺（交联剂）和直径

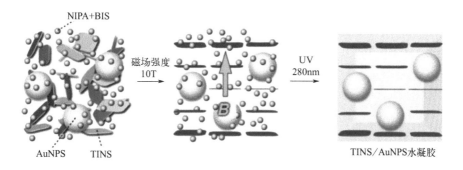

图9.18 基于紫外光响应形变的TINS/AuNPS水凝胶执行器的示意图[103]

为 17nm 的 TINS（光响应剂）均匀分散在 N- 异丙基丙烯酰胺（高分子单体）溶液中自发聚合而得。该水凝胶执行器能对波长为 445nm（5.6W/cm² 电源密度）的光展现出高效的光热转换性能。因此，在紫外光照射后，水凝胶表面温度可以快速升高到 85℃，该温度下水凝胶基体开始膨胀变长从而实现需要的变形。当关闭激光时，水凝胶在 LCST 以下自然冷却并在 6s 内收缩回原来的长度；当反复打开和关闭激光灯时，水凝胶基体的重复性拉长和收缩可以让水凝胶执行器实现动态的爬行过程。

9.2.2.3　近红外光响应水凝胶执行器

近红外光由于其易于获取以及高能量等特性也被广泛用于响应性水凝胶执行器的构建之中。Xue 等[104]报道了一种具有近红外光驱动的程序化变形能力的 MXene 水凝胶执行器，由可光聚合的 MXene 纳米单体和 PNIPAM 水凝胶之间的原位自由基共聚反应制备而成（图 9.19）。其中，结构化水凝胶的厚度可以通过电场调节 MXene 纳米片的浓度梯度来控制。基于该机理，他们通过施加图案化电场光聚合，获得了一系列几何形状的各向异性水凝胶执行器，包括 "U 形""J 形""S 形""W 形"乃至"花形"水凝胶驱动器。其中，"花形"各向异性水凝胶执行器可以实现精确的变形驱动过程，其演变行为类似于真花的开花闭合行为。此外，该水凝胶执行器还可作为四臂软夹具，在程序化的近红外光刺激下实现抓取、提升、降低和释放物体等一系列复杂的操作。这种方法的吸引力在于它促进了近红外光刺激的更广泛应用。以上这些光控策略可以为开发更智能的水凝胶执行器或机器的开发提供新见解，赋予执行器具有更复杂且可重新配置的动作，以实现新颖的多功能应用。尽管已经有很多不同领域的光响应水凝胶的研发，但当前的大多数系统都仅限于简单的应用研究。为了扩大其实际应用，需要解决一些主要的挑战，包括开发光化学反应、光动力学的研究以及光响应水凝胶的生物相容性等。尽管需要解决许多挑战，但光响应水凝胶执行器的发展和实施的前途是光明的。

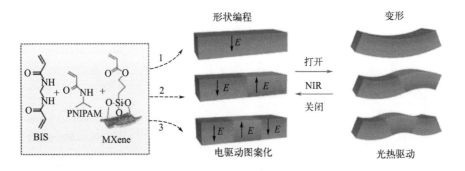

图 9.19　近红外光响应的 MXene 水凝胶执行器的示意图[104]

9.2.3 磁响应

不同于光/热响应执行器，在微创手术和靶向药物输送等特定的应用场景下，磁响应执行器更为实用。因为它不仅可以在短时间内被磁场远程触发，而且在高磁场通量中也能保持生物相容性。该类水凝胶执行器的响应行为主要是通过在高分子基质中加入外部顺磁性或铁磁性添加剂来实现，这种磁场刺激响应的特点是快速、大形变驱动以及可远程控制（图 9.20）。根据磁化特性，磁性添加剂可以分为两类：软磁性（包括金属铁[105]、合金与氧化物[106,107]，该类物质具有高磁化率和饱和磁化等特征）和硬磁性［如钕铁硼磁体（NDFeB），最典型的特征是具有较大的磁性磁滞，因此可以作为永久磁场］。这种磁响应水凝胶执行器在合成过程中一般不需要额外的交联剂，磁性颗粒会在凝胶交联的过程中自发地以共价或协同作用的形式与凝胶中的高分子形成交联。接下来我们将具体介绍以 Fe_3O_4 颗粒和 NDFeB 作为主要的磁性掺杂剂的磁响应水凝胶执行器的合成和驱动过程。

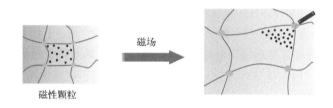

图 9.20　磁响应的水凝胶致动机理的示意图[86]

9.2.3.1　Fe_3O_4 颗粒响应水凝胶执行器

磁性添加剂一般是直接通过物理作用嵌入水凝胶网络中，在受到外部梯度磁场的作用时，水凝胶基质中的磁性添加剂充当磁性接收器并可受到外部非均匀磁场力的操控。具体表现为磁性添加剂向前或向后运动，最终形成与外部磁场一致的梯度分布，进而驱动水凝胶结构的变形和运动。Li 等[108]制备了一种兼具 pH 响应和磁性驱动的双层响应体系的水凝胶执行器（图 9.21），由甲基丙烯酸 2-羟乙酯（PHEMA）层和掺杂 Fe_3O_4 颗粒的聚乙二醇二丙烯酸酯（PEGDA-Fe_3O_4）层组成，其中 PHEMA 层具有 pH 响应特性，而 PEGDA-Fe_3O_4 层可以响应外部磁场的变化。因而该双响应水凝胶执行器可以作为微型机器人在不同 pH 值的溶液以及电磁驱动系统的辅助下实现药物的捕获和释放。具体驱动过程如下：柔软的微型机器人在 pH 值为 9.58 处抓取抗癌药珠，并在电磁驱动系统作用下软性微型机器人会转移到对应肿瘤组织；在瞄准肿瘤后，软性微型机器人可以精准释放之前捕获的药珠，因为肿瘤组织的 pH 值约为 2.6，这低于正常组织。综上，这种软微型机器人可以通过电磁驱动系统移动到目标位置，并且配合溶液环境的 pH 变化实现药物在特定条件下的抓取和释放。若进一步对外部磁场和执行器程序化配置后，复合软体执行器就可以在各种复杂地形上实现多模式运动。

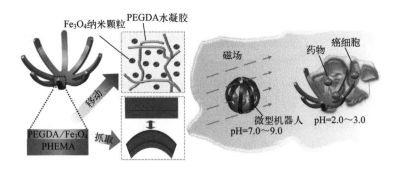

图 9.21　双磁调控水凝胶层的示意图[108]

Podstawczyk 等[109]以结合了磁性纳米粒子的 PAA 分散溶液与海藻酸盐和甲基纤维素水凝胶的混合前体为打印浆料，并通过 3D 打印技术构建磁性水凝胶执行器。其中，水凝胶中嵌入的 PAA 与磁性纳米粒子起到交联剂的作用，它们与海藻酸盐和甲基纤维素等高分子骨架通过静电相互作用连接，磁性纳米粒子的掺入使得打印的水凝胶微结构可以在受到外部磁场的作用时发生变形。在施加磁场（MF）的情况下，水凝胶条快速向磁场方向靠近并实现最大程度的弯曲（图 9.22），这种最大弯曲程度可以通过加入磁颗粒的密集程度决定，可以实现弯曲角度从 24°～106°。磁性水凝胶执行器可以通过外部磁场的改变精准地控制柔性材料膨胀程度的特点丰富了新型刺激响应式药物输送系统——可以将药物负载到水凝胶的网格结构中，通过改变外在磁场来控制水凝胶网格结构的大小来控制药物的释放。

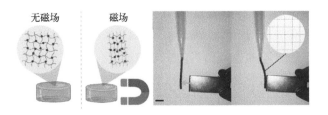

图 9.22　磁响应水凝胶执行器的示意图（比例尺为 2cm）[109]

9.2.3.2　NDFeB 响应水凝胶执行器

相比于软磁性水凝胶执行器较低的编程灵活性，基于硬磁材料 NDFeB 构建的磁响应水凝胶执行器往往具有良好的可编程性并且可以实现更复杂的形状转化。当将 NDFeB 颗粒掺入软高分子基质中时，所得的软体执行器在表现出硬磁性特征的同时（即高透明和高矫正性），还具有机械柔软和可灵活变形等特性，因此展现出优异的应用前景和价值。Cheng 等[110]报道了一种自修复超分子磁弹性体（SHSME），由于弹性体内部强大的动态网络和丰富的可逆化学键，SHSME 展现出较高的机械强度（弹性模量为 1.2MPa）和快速的自愈能力（在环境温度下修复 5s 后拉伸应变达到 300%）。基于两个磁化 SHSME 片通过可逆化学键可实现快速界面键合，能够有效调控 SHSME 内部的局部结构磁化分布。由三个短磁化 SHSME 单元组成的 SHSME 棒，受到均匀外界磁场的作用后，会经历快速和可预测的形状转变过程（图 9.23）。这种磁控软驱动的功能是可逆的，当撤去磁场后，水凝胶驱动器又会立马恢复成棒状。值得注意的是，SHSME 框架不仅可以利用热变形回收并重复使用，还可以实现模块化定制结构和功能。上述结果表明这种 SHSME 软体机器人可以实现复杂的几何形状以及多功能的模块化组装，有望应用于限定空间内的肢解 - 导航 - 组装等策略任务。

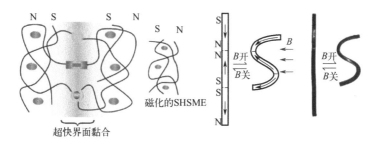

图 9.23 磁化 NDFeB 水凝胶执行器的示意图[110]

9.2.4 电响应

电场响应水凝胶执行器，也称为电解质或电活性水凝胶执行器，这类执行器在受到溶剂诱导或外部施加的电场的作用时会通过膨胀或收缩等行为响应外部环境的变化[111]。使用电场作为外部刺激具有独特的优势，即可以通过软件编程实现对电压 / 电流的精确控制，从而精准操控水凝胶执行器的运动和行为。电响应水凝胶执行器的主要工作机制通常取决于电场的离子运动以及水凝胶的溶胀介质内离子的分布 / 排列（图 9.24），这些参数都可以通过电极的相对位置（如单体浓度和孔径大小）以及水凝胶的溶剂特性（如离子强度）来控制。具体来说，当受到外部电场的作用时，水凝胶内部的自由离子会在介质中发生定向迁移，使得水凝胶内外的离子浓度分布不均匀并产生渗透压，从而驱动水凝胶执行器溶胀或收缩[112]。基于电场控制快速、精准和重复驱动等特点，电响应水凝胶执行器可以在生物工程和生物医学等应用中展现出重要的应用前景。现有研究中主要存在两种典型的电响应水凝胶执行器（直流电响应和交流电响应水凝胶执行器），由于两者的外部控制电流源存在区别，因而将两者分开论述和说明。

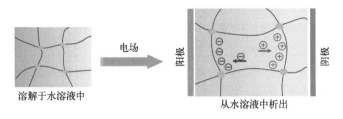

图 9.24 电响应水凝胶执行器的致动机理示意图[86]

9.2.4.1 直流电响应水凝胶执行器

由于大部分小型电源是以直流电信号的方式输出的，因此，经常可以在一些小型化/微型化的设备中见到直流电响应的水凝胶执行器。其中，聚2-丙烯酰胺-2-甲基丙磺酸（PAMPS）是构建直流电响应水凝胶执行器最受欢迎的成分之一，因为该单体可以在pH=2～12的宽范围水溶液中完全电离，从而赋予PAMPS水凝胶执行器良好的直流电响应性能。Yang等[113]报道了一种具有出色的直流电响应和力学性能的新型还原氧化石墨烯/聚2-丙烯酰胺-2-甲基丙甲磺酸-丙烯酰胺[rGO/Poly（PAMPS-co-AM）]复合水凝胶执行器。该水凝胶执行器中均匀分散的还原氧化石墨烯纳米片使其具备良好的导电性，这有利于促进水凝胶内部的离子传输并调控水凝胶内部/外部的渗透压（图9.25）。此外，rGO纳米片和高分子链之间的氢键相互作用进一步提升了水凝胶的力学性能和抗压强度，从而赋予这种水凝胶执行器优异和快速的可逆变形能力。在电解质溶液中，rGO-0.5（其中"0.5"代表纳米复合水凝胶中rGO的质量浓度为0.5mg/mL）水凝胶执行器可以在90s的时间就可以提起0.4g的重物（图9.25）。该结果表明，rGO纳米片的掺入不仅显著提高了水凝胶的电响应性能和力学性能，同时还提升了它的致动性。尽管这种水凝胶执行器展现出了高变形能力和快速响应时间，但是，体积过大以及需要在溶液中才能进行驱动等问题限制了它的应用。

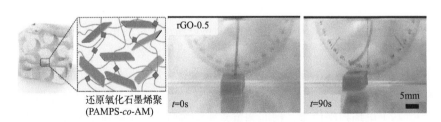

图 9.25 直流电调控水凝胶执行器的示意图[113]

Kang 等[114]通过 3D 打印的方式构建了一种基于聚 3-磺基甲基丙烯酸酯（PSPA）的直流电水凝胶执行器，基于该方式制备出来的执行器通常具有可定制化的结构和体积。水凝胶执行器被置于 KCl 溶液中的两个电极之间（图 9.26），当外电路闭合时，由于电场的作用，水凝胶两侧界面的离子浓度会出现差值，进而在界面上产生电渗透压。为了实现水凝胶的电中性，电解质溶液中的阴离子会迁移到水凝胶中，同时水凝胶的阳离子迁移到电解质溶液中，最终水凝胶执行器在电场力的作用下表现为向负极弯曲（图 9.26）。尽管该水凝胶执行器在 7.5V 直流电的驱动下产生的离子传输速度仅为 1mm/min，需要经历 10min 才能完全驱动水凝胶样品变形；但是在 7.5V，0.05mol/L KCl 溶度条件下，水凝胶执行器可以在 4min 内实现弯曲驱动（图 9.26），这是因为水凝胶在 0.05mol/L KCl 溶液中具有最佳的电活性 C_{KCl}。上述结果表明了该直流电水凝胶执行器具有可调可控的特性。此外，这种电响应水凝胶的形状变化与天然的肌肉组织相似，因此可以用作生物传感器和人造肌肉。然而，由于直流电电压一般较小，会使得相应的驱动速度较慢，这限制了其进一步的发展和应用。

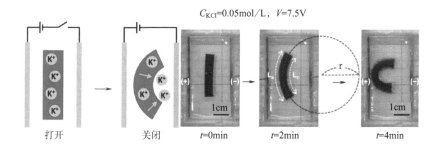

图 9.26　直流电响应水凝胶执行器的示意图[114]

9.2.4.2　交流电响应水凝胶执行器

交流电响应水凝胶执行器由于其更加快速高效的驱动能力，因此同样得到了深入的开发和研究。Ko 等[115] 利用丙烯酸基水凝胶作为多孔膜和离子液体的储存介质，制备了一种电压渗流和膨胀驱动的水凝胶执行器，该水凝胶执行器具有低工作电压、快速驱动以及可以实现多自由度运动等特点（图 9.27）。当向开路的两端电极施加大约 3V 的交流工作电压时，水凝胶内部储存的大量液压流体开始迅速流动，导致水凝胶发生膨胀变形，从而实现驱动。这种水凝胶执行器主要由受交流电压激发的水合离子液体膨胀驱动，而且可通过调节交流电压改变运动的幅度，为开发和设计各种可控软电子设备提供了思路。尽管该材料展现出在肌肉重建领域的巨大应用潜力，但是，由于需要连接电线等外电路设备才能实现电驱动,整个装置过于冗杂笨重,因此需要进一步改善。

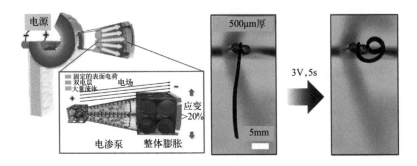

图 9.27 交流电响应水凝胶执行器的示意图[115]

除了上述的各种物理刺激之外,化学刺激同样可以实现对水凝胶执行器的可控驱动。在此,以两种最常见的化学刺激手段进行介绍,包括 pH 响应和离子响应水凝胶执行器。

9.2.5 pH 响应

众所周知,pH 响应水凝胶执行器是通过体系中的 pH 变化来影响水凝胶网络上的渗透压差而实现响应驱动。大多数 pH 响应水凝胶执行器在其高分子骨架上含有可在溶液中电离的侧基,根据电离后的电荷类型,可分为阴离子侧基和阳离子侧基两大类。其中,阴离子侧基水凝胶执行器具有在 pH 高于其酸解离度(pK_a)时解离侧链基团的特征,当侧链基团解离后,水凝胶会由于失去质子而带负电,水凝胶网络内部渗透压减小,从而在渗透压的驱动下使水凝胶膨胀变形。当 pH 低于 pK_a 时,阴离子侧基水凝胶则会收缩。与之相反,阳离子侧基水凝胶在 pH 低于 pK_a 时发生膨胀,高于 pK_a 时出现收缩(图 9.28)[116,117]。

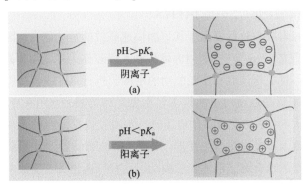

图 9.28 pH 响应水凝胶执行器的致动机制示意图,包括阴离子响应性(a)和阳离子响应性(b)[86]

9.2.5.1 阴离子侧基响应水凝胶执行器

Duan 等 [118] 合成了一种基于天然纤维素 [左边为壳聚糖（CS）水凝胶层，右边为纤维素/羧甲基纤维素（CMC）水凝胶层] 的阴离子侧基响应双层水凝胶执行器（图 9.29）。在相对高 pH 条件下（pH > 3.8），水凝胶执行器的 CMC 层侧基中大量羧基解离，使得水凝胶层失去质子带负电，进而导致双层水凝胶向 CS 层的一侧弯曲。相反，在低 pH 下（pH < 3.8），由于 CS 层的溶胀率高，氨基质子化，此时双层水凝胶的 CS 层向另一方向弯曲。当 pH=3.8 时，CS 水凝胶层与 CMC 水凝胶层的溶胀情况一致，因此处于不偏向任何一侧的稳定状态（竖直状态）。基于上述特性，他们进一步构建了一系列可以通过 pH 触发的具有机械工作能力的新型软执行器，该类执行器不仅可以实现可逆和反复的可控变形，还实现包括"环""花""螺旋竹子""胶囊"等各种复杂的形状变化，在软抓手、智能封装器和人工肌肉等领域展现出潜在的应用价值。

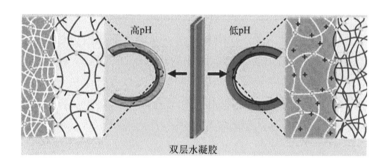

图 9.29　阴离子侧基 pH 调控的双层水凝胶执行器示意图 [118]

9.2.5.2 阳离子侧基响应水凝胶执行器

Li 等 [119] 使用 pH 响应的聚丙烯酸水凝胶微管，制备了一种阳离子侧基水凝胶执行器。当溶液的 pH 在特定范围内发生转变时，由于可以响应 pH 并实现不对称的膨胀/收缩，该微管可以在 1s 内呈现出可逆的弯曲和直立状态的动态演变过程。当溶液的 pH 值小于 9 时，丙烯酸侧链的羧基响应溶液 pH 变化而实现质子化，微管由直立状态快速收缩成弯曲状态；当溶液 pH 值小于 9 时，微管恢复直立状态（图 9.30）。此外，进一步通过一种动态全息处理和拼接方法，可以在大约 10s 内生成更复杂的 pH 响应微结构（包括"S 形""C 形"微管和手性扭转结构）。这种微型水凝胶执行管可以用于捕获和释放聚苯乙烯颗粒和神经干细胞。凭借优异的 pH 响应性，该类阳离子侧基水凝胶执行器已被广泛用作生物传感器和药物递送载体。然而，在实际应用中，必须克服在干燥条件下无法控制的开/关驱动状态和控制过程费时的缺点。

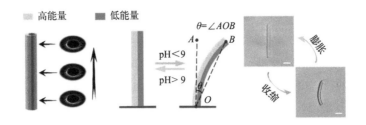

图 9.30　阳离子侧基 pH 响应水凝胶微结构执行器示意图 [119]

9.2.6　离子响应

离子响应水凝胶执行器，可以随着溶液离子种类或浓度的变化而膨胀和收缩，为水凝胶执行器在水下的应用提供了更多的可能性。通常，许多 pH 响应水凝胶也是离子响应水凝胶，其内部聚电解质网络在纯水中膨胀而在溶液中收缩，主要是由于水凝胶基质内渗透压的变化。在低离子溶液中，由于溶液中可电离基团的解离，水凝胶在主链上产生高电荷密度，从而在基质内引起分子间排斥，进而使得分子间距扩大，最后表现为水凝胶的体积膨胀 [120]。一般情况下，离子响应水凝胶由两性离子高分子组成，其离子响应性可以通过添加不同类型和浓度的离子来调节 [121]。接下来，我们将以 Fe^{3+} 和 Ca^{2+} 响应水凝胶执行器为代表分析和讨论离子响应水凝胶执行器的驱动行为。

9.2.6.1　Fe^{3+} 响应水凝胶执行器

一般情况下，Fe^{3+} 可以特异性结合羟基等化学基团以产生驱动行为。Le 等 [122] 利用聚丙烯酸与 Fe^{3+} 形成聚丙烯酸 -Fe^{3+} 螯合物，赋予整个水凝胶材料的各向异性结构具有离子驱动的行为。当改变滤纸中 Fe^{3+} 浓度时，水凝胶膜在 1min 内随之呈现不同程度的弯曲度（图 9.31）。这种控制方式同样可以用于控制三维"花"

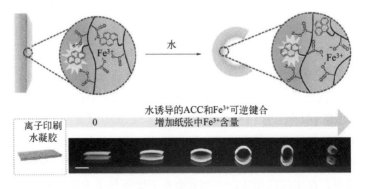

图 9.31　Fe^{3+} 调控的水凝胶执行器（比例尺为 1cm）[122]

形水凝胶的"开花"程度。主要是因为当水凝胶与含 Fe^{3+} 溶液接触时，Fe^{3+} 进入水凝胶与聚丙烯酸发生三齿螯合交联，降低水凝胶一侧的溶胀率，从而导致水凝胶的弯曲行为。当增加 Fe^{3+} 含量时，由于水凝胶内部产生更多的螯合交联以及各向异性结构，水凝胶执行器展现出更明显的弯曲变形。此外，通过调节 Fe^{3+} 含量，还可以智能调节复合水凝胶执行器的荧光性能，这种荧光水凝胶在离子溶液中可以实现信息隐藏与展现，在信息存储方面具有巨大的应用潜力。

9.2.6.2 Ca^{2+} 响应水凝胶执行器

Du 等[123]通过光刻获得图案化的微通道，并在微通道中填充了预凝胶溶液，随后将图案化预凝胶浸入 $CaCl_2$ 溶液中形成离子交联，剥离得到水凝胶薄膜执行器（图 9.32）。将水凝胶片沿不同方向切割成微通道并浸入 $CaCl_2$ 溶液中，以实现微通道中水凝胶结构的可控 3D 结构转换。将所得的水凝胶膜片可以做成复杂的花瓣状，并用 0.1mol/L Ca^{2+} 处理"六瓣花"结构水凝胶，其中五个"花瓣"部分涂有蓝色染料，可以完整模拟花朵的开合过程，其他"花瓣"在非微通道表面写有字母"R"。可以清楚地看到，随着控制时间的推移，浸泡 3min 后带有字母"R"的闭合"花瓣"慢慢张开。7min 后，它完全打开成一朵"花"。将"六瓣花"浸泡在 0.1mol/L $CaCl_2$ 溶液中，浸泡 9min 后，"花朵"闭合成带有字母"R"的完整"花蕾"。以上整个过程可以生动地展示"开花"的不同阶段，并且字母"R"在"花瓣"上的标记可以记录水凝胶的向内向外可逆变形过程。该方法为未来高分子形状控制提供了新的可能和指导，并且可以进一步控制变形参数来实现高精度的图案化和控制响应过程，展示出极大的研究和应用价值。

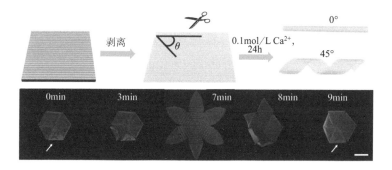

图 9.32 Ca^{2+} 响应的水凝胶执行器的示意图（比例尺为 1cm）[123]

9.2.7 湿度响应

自然界给了我们很多软体机器的控制思路和方式，而空气中的湿度作为刺激响应的常见手段被广泛研究。松果可以在空气干燥时打开其鳞片，在潮湿时关闭；小麦也表现出类似的功能来实现种子的释放过程。湿度是一种易于实现的控制策略，因此具有湿度响应的水凝胶执行器在未来的各种工程应用中具有巨大的潜力[124,125]。

为了探索湿度响应水凝胶执行器的功能，Sun 等[126] 利用高吸湿性的聚乙二醇二丙烯酸酯（PEG-DA）为主链，添加光敏剂后，PEG-DA 单体很容易聚合，结合 3D 打印制备了具有优异湿度响应的 3D 微结构水凝胶执行器，并表现出良好的形态结构和稳定的"花朵"状；当把这种打印好的水凝胶"花"浸入水中或将周围的环境湿度增加时，水分子扩散到高吸湿性的 PEG-DA 水凝胶分子网络中以扩大内部微观结构，表现为吸水后的"花"在几秒钟内明显变大。吸水后的"花"的直径扩大至原本的 155%，而从水中取出后，"花"逐渐缩回原来的状态，这个过程可重复多次（图 9.33）。除了将"花"直接放入水中，增加环境湿度也会使凝胶膨胀，这种溶胀过程可以重复 10000 次，整个水凝胶微观结构在重复过程中具有良好的稳定性。这种复合水凝胶微观结构可能会在执行器或软机器人的构造中带来更广泛的应用。

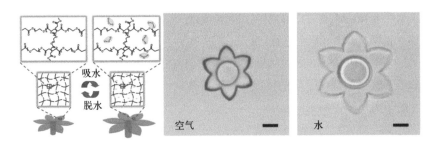

图 9.33　基于 3D 打印的可编程水凝胶湿度响应性执行器的湿度响应
机制示意图（比例尺为 10μm）[126]

Zheng 等[127] 利用聚乙烯醇与 9-蒽酸接枝，并与紫外线屏蔽剂 2, 2', 4, 4'-四羟基二苯甲酮（THBP）共混，交联制成了一种湿响应水凝胶执行器。这种水凝胶执行器可以根据需要编程和调控复杂的形状变化。其中，蒽基团可以通过控制紫外线感应使执行器进行复杂图案化和可逆编程控制。由于交联梯度是通过添加 THBP 构建的，随着外部湿度的变化，水凝胶的内部膨胀随着交联网络的密度而变化，从而产生变形行为。这种水凝胶执行器在加热水蒸气的湿度场中的执行性能的变化，当环境相对湿度从 20% 上升到 90% 时，执行器迅速向紫外辐照侧弯曲，在只有紫外线辐射时水凝胶执行器不弯曲（图 9.34）。利用该水凝胶执行器设计的一个柔软的步行机器人，可以在湿度的间隔刺激下向上拱起并直接向下弯曲，在弯曲过程中，由于前边缘和后边缘的接触区域摩擦力的差异，前缘带有较大的摩擦将充当固定端，而后端将向前移动，从而实现了复杂的控制运动行为。

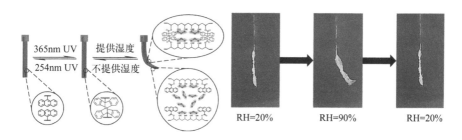

图 9.34 聚乙烯醇梯度交联湿度响应性执行器的示意图[127]

9.2.8 溶剂响应

由于水凝胶类似于人体结构并且内部含有大量的水而实现了许多功能应用。当在一些有机溶剂存在的情况下，由于溶剂和水凝胶内部水分和离子等的相互作用，从而表现出一些相应的致动和记忆行为。以下将聚焦到乙二醇和丙酮等溶剂对于水凝胶执行器作用功能和行为的影响。

9.2.8.1 乙二醇响应水凝胶执行器

Wu 等[128]实现了一种可以响应水和乙二醇的智能水凝胶执行器。他们利用乙烯基功能化碳量子点结合丙烯酰胺（AM）和丙烯酸（AAc）自由基共聚制备蓝色荧光量子点高分子凝胶层，随后将 3M VHB 弹性体层与凝胶层贴合获得非对称双层执行器。该执行器在水溶液中由于化学交联的聚（AAc-AM）凝胶的羧基将在高介电常数水中部分电离，导致高分子链之间产生静电排斥而表现出水凝胶层吸水膨胀，引起弹性体层变形。平展的双层执行器条纹在水中 60min 内逐渐扭曲成弯曲的条纹，在乙二醇中又逐渐恢复到原状（图 9.35）；在这个过程中，

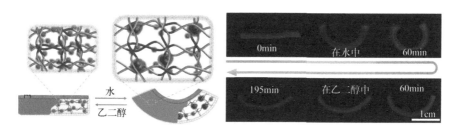

图 9.35 灵感来自章鱼的乙二醇响应性碳量子点水凝胶的示意图[128]

伴随着显著的荧光颜色变化（从深蓝色变为玫瑰色）。这种形状和颜色的变化高度同步，整个过程都由溶液扩散控制并由相同的刺激（即水 - 乙二醇交换）触发。因此，这种智能荧光水凝胶执行器可用于配合形状变形、颜色变化和定向运动的游泳水凝胶机器人，用荧光变化作为应对外界危险时的警报提醒功能，综上，这种水凝胶执行器的协同变色 / 变异行为可以很好地激发仿生变色的发展和未来的多功能软机器人。

9.2.8.2 丙酮响应水凝胶执行器

除了以上提到的乙二醇响应水凝胶执行器，这里还将介绍丙酮响应水凝胶执行器。Hubbard 等[129] 报道了一种使用玻璃纤维界面将水凝胶与弹性体结合的方法，该方法可以实现对复合织物的溶剂控制效果。他们制备了三层弹性体 / 织物 / 水凝胶层，利用织物与柔软的弹性体的相互作用，通过静电相互作用与玻璃表面相互作用和附着力将水凝胶填充在中间（图 9.36），复合层中的水凝胶层可以产生溶剂溶胀响应。由于溶胀过程缓慢，通过调节弹性体加入比例来降低弹性层的模量，从而使复合材料更容易弯曲变形。复合材料在去离子水中保持铺展状态，在丙酮或 2.0mol/L 盐溶液之间交替以出现相反方向弯曲变形。这种复合执行器为更复杂的三态或多态执行器提供了新思路。

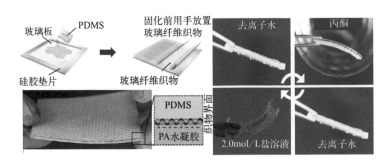

图 9.36　通过织物界面黏结的水凝胶 / 弹性体层压板的丙酮响应性执行器[129]

溶剂响应水凝胶执行器同样有着十分广泛的应用场景和独特的控制方式，但是这种方式也有着一些限制，比如响应不够快速，受外界影响严重和溶剂挥发造成的环境问题等。

9.2.9　其他响应

除了上述各种刺激方式之外，其他类型的物理或化学刺激，如液压和生物分子同样也被用于触发水凝胶执行器。一般情况下，高分子网络的独特设计可以实现可逆和可控的形状转变功能。下面给出几个例子来讨论这两种刺激手段控制的水凝胶执行器变形行为。

9.2.9.1 液压响应水凝胶执行器

液压响应水凝胶执行器根据外部液压变化进行形状调控,与其他现有水凝胶执行器相比,它们可以产生更高的执行力。Yuk 等 [22] 利用坚韧水凝胶中的物理和化学混合交联以及交联耗散网络制备了基于聚丙烯酰胺(PAM)和海藻酸盐的液压水凝胶执行器和机器人,需要将得到的水凝胶执行器浸入水中 48h 以达到平衡状态。此外,他们还进一步使用 3D 打印构建具有复杂结构的水凝胶执行器,获得的水凝胶执行器能够维持多次液压驱动循环。如图 9.37 所示,这种水凝胶执行器的驱动行为是由施加在外部的液压产生的,这种驱动力可以达到大于 1N 的驱动变形力,以及更快的响应速度(响应时间小于 1s)。此外,这种水凝胶表现出高透明度,因此可以在水中进行光学和声学伪装作为水下钓手,可以成功捕捉和释放水中金鱼,为水凝胶执行器应用于海洋生物学中的下一代仿生机器人提供新的思路。

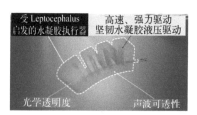

图 9.37 对水压响应的水凝胶执行器和机器人 [22]

9.2.9.2 生物分子响应水凝胶执行器

生物分子响应水凝胶代表了现代生物材料的重要前沿 [130]。最具有代表性的生物分子响应水凝胶是 DNA 响应水凝胶,然而,DNA 响应水凝胶的体积膨胀率很小(在 10% ~ 20% 范围),这明显小于 pH 或温度响应的体积转变,因此限制了大规模和高精度的执行变形行为和运动。为了解决这种局限,Cangialosi 等 [131] 提出了一种由 DNA 发夹组成的水凝胶执行器,他们通过依次添加不同的 DNA 发夹,从而改变水凝胶的体积扩展变形,其体积变化可以达到原来 DNA 响应水凝胶的 100 倍(图 9.38)。并进一步探索了几种不同 DNA 序列的结构对于水凝胶执行器形状的不同控制过程,例如,他们制备了一种水凝胶"花朵",其中一组"花瓣"的 DNA 序列不同于另外两组,从而在不同 DNA 刺激的情况下可以展现出不同的变形反应。在两组不同 DNA 序列的情况下,所有的"花瓣"都可以折叠起来,每组序列可以实现对于"花瓣"的单独调控折叠行为,从而实现复杂的"花瓣"可以逐步折叠过程。这种驱动过程也可以用于更复杂的结构如触角、爪子和腿,可以根据它们在水凝胶装置上的各自顺序同时或顺序地卷曲和执行。这为新型生物分子响应执行器提供了新的方向。

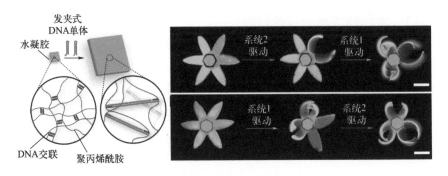

图 9.38　DNA 响应水凝胶执行器的示意图（比例尺为 1mm）[131]

9.3　仿生智能水凝胶执行器的应用

　　水凝胶执行器的潜在应用与其可逆的形状转换性能及刺激响应方式密切相关。本章的前两节对于各种仿生响应水凝胶执行器的响应和执行行为的控制方式进行了介绍，并简单涉及了一些可能的应用场景。随着材料设计和制造技术的进步，软执行器和设备或机器人的发展越来越复杂，已在多个领域应用。尽管很多应用仍处于概念阶段，但水凝胶执行器已应用于电子皮肤、人造肌肉、软执行器、软机器人、生物传感器、药物输送、组织工程和夹具（图 9.39）[132-134]。在本节中，我们将介绍水凝胶执行器的各种潜在的现实应用，从响应软执行器和流体控制到医学工程。

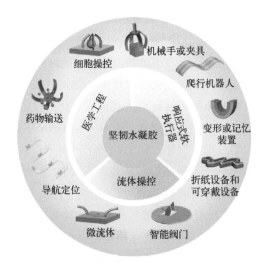

图 9.39　水凝胶执行器潜在应用的示意图

9.3.1 软执行器

9.3.1.1 机械手或夹具

水凝胶执行器最有前途的应用之一是作为相对复杂的多臂机器人或夹持器。当暴露于一种或多种刺激时,水凝胶执行器会捕捉物体,并在另一种刺激下将其释放。尽管这一应用仍处于初始阶段,但刺激响应执行器已被广泛验证为智能执行器或机械手、柔性传感器、移动医疗和外科活检工具等,这些都与其形状编程密切相关[133]。软夹持器,作为一种广泛应用的执行器,可根据特定用途选择系统实现刺激抓取和响应性释放的过程。基于这一机理,Yuk 等[22] 开发了一种光学和声学伪装的液压软透明夹具,这种夹具可以在没有限定区域的情况下抓取和释放活金鱼(图 9.40)。他们在水中使用具有高光学和声学透明度的水凝胶执行器,以实现类似于光学和声波环境中纤维的自然伪装,这种液压水凝胶执行器中的高含水量使其具有与水几乎相同的视觉和声波特性,从而达到自然的伪装,液压驱动的水凝胶驱动器可以实现类似于鱼的运动(图 9.40)。因此伪装的水凝胶启动器可以捕捉和释放水中的活鱼,当水凝胶紧贴水箱中的金鱼时,其光学透明度保持其伪装状态,灵活的驱动控制可以成功捕获金鱼,内部柔性在释放捕获的金鱼时不会对金鱼造成任何伤害。这种水下活体机械手夹具为相关研究开辟了道路,但是这种夹具只能通过单一刺激控制,从而限制了其实现更复杂的应用。

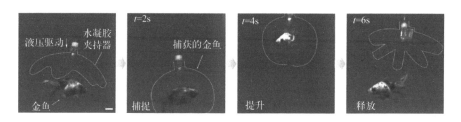

图 9.40 水凝胶执行器作为操纵器或抓手的功能;捕捉金鱼使用的水凝胶抓手图像(比例尺为 1cm)[22]

Dong 等[135] 采用简单但稳健有效的方法,通过添加氧化石墨烯(GO)/聚吡咯(PPY)来实现可编程的图案化 GO/PPY 双响应水凝胶执行器,此执行器表现出在温度和红外光的刺激下优异的响应执行性能。受鹰爪的启发,他们设计了一个带有十字结构的智能抓手(图 9.41),类似于老鹰可以用爪子捕捉猎物,GO/PPY 夹具可以在高湿度刺激下弯曲并成功拾取重量为自身 38 倍的方形聚合泡沫,在低湿度时释放重物并恢复到初始状态,表现出优异的机械夹持特性,为制造可以在许多领域工作的水凝胶机器人提供了一种有效的方法。

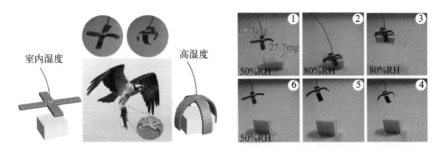

图 9.41　GO/PPY 水凝胶仿制鹰爪的抓手照片[135]

类似地，Yao 等[136] 开发了一种湿度响应（水合 / 脱水）水凝胶执行器，该执行器由 Eu^{3+} 型聚乙烯亚胺共聚丙烯酸（PEI-*co*-PAA）以及聚乙二醇二丙烯酸酯（PEGDA）组成，这种水凝胶执行器表现出环境湿度响应的水合 / 脱水敏感的特性，并在紫外光刺激下实现物体定位和识别的功能。利用以上特性设计和制备了伪装水母的双层六角水凝胶，用于实现水下物体的抓取和释放过程（图 9.42），并进一步探索了水下的适用性，发现这种水凝胶抓手可以抓取直径约 4.5cm，重约 25g 的物体，因此这种动态伪装水凝胶可以用来捕捉活的金鱼。在打开状态下，抓手在日光下不透明，在抓住金鱼后，经紫外线照射可发光，从而很容易定位；该执行器丰富了水下软体抓手的更多可能性和操作场景。

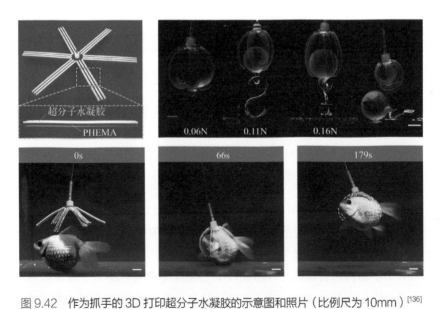

图 9.42　作为抓手的 3D 打印超分子水凝胶的示意图和照片（比例尺为 10mm）[136]

以上提到的执行器用于抓取和释放动作时都需要将一端固定，再实现对于目标物体的抓取和释放的功能，因此会存在较大的功能灵活性的限制，为了摆脱这种限制，随后开发出了另一种类型未固定的软执行器，这种不受约束的小型软机器人可以在复杂环境下中提供更多的抓取释放可能。Zhao 等[137]报道了一种 4D 打印制备的具有双层不对称结构的水凝胶执行器，打印水凝胶墨水中含有纳米型囊泡、合成锂皂石、N-异丙基丙烯酰胺、N,N'-亚甲基双丙烯酰胺和苯锂-2,4,6-三甲基-苄基膦酸酯，其中纳米型囊泡可以通过光热催化分解 H_2O_2 释放 O_2，合成锂皂石用于调控打印质量，因此这种水凝胶执行器具有 O_2 释放的能力。随后，将打印好的水凝胶片组装成四臂水凝胶执行器实现物体转移和释放的过程（图 9.43），水凝胶执行器在 50℃的水中逐渐闭合，随着浸泡时间增加，执行器捕获连接在绳索上的橡胶块，然后将块体紧紧包裹，通过拉绳运动可以实现移动包裹的过程。此外，这种水凝胶执行器的弯曲/释放触发器形状替换为仿生植物花和水凝胶机械手，这两种结构同样在热刺激下很容易实现目标物体的关闭/打开和抓取/捕获的操作。基于这种思路开发了各种更复杂的水凝胶执行器可用作更智能的夹具，并提供更多机会来克服苛刻环境条件下夹具的使用限制[138,139]。

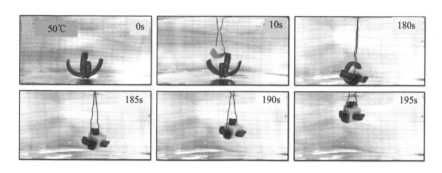

图 9.43　4D 打印制备的水凝胶执行器作为抓手的示例[137]

9.3.1.2　爬行器或步行器

周期性的刺激、弯曲动作可以转化为爬行或行走的表现，从而在更丰富的实际场景中应用。受蚯蚓运动的启发，Sun 等[103]开发了一种圆柱形各向异性爬行水凝胶执行器，由热响应高分子网络和二维纳米电解质在外部光源的刺激下控制水凝胶的单向膨胀和收缩从而实现执行器的爬行行为。当使用 445nm 激光束照射右端的水凝胶时，照射区域逐渐变薄，从而减少与毛细管壁的摩擦，水凝胶执行器的右边微微抬起；当照射点向左端移动，对应区域的水凝胶逐渐延伸/变薄，与接触面产生摩擦，从而实现类似于蚯蚓的爬行运动，整个水凝胶柱实现从左向右移动（图 9.44）。当重复反向操作时，水凝胶柱也可以实现方向转变的爬行过程。当把内部的金纳米粒子（光热剂）替换为金纳米棒时，转变为近红外光远程控制水凝胶，由于近红外光可以穿透生物组织，表明该系统可以实现通过外部控制在体内发挥作用的功能，以应用到临床药物递送和伤口检测等场合。

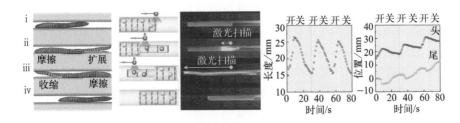

图 9.44　光响应性类蚯蚓水凝胶执行器作为爬行或行走的功能的示意图[103]

同样尺蠖的爬行过程也被广泛研究和模仿，它的运动过程可以分为两个步骤（图 9.45）：①前脚紧紧抓住树枝的同时身体蜷缩；②抓住树枝后，身体前段向前移动从而带动整个身体的前移。受此启发，Dong 等[135]利用氧化石墨烯（GO）和聚吡咯（PPY）制备了一种双层水凝胶执行器，一部分有水平面和角度调节的能力（A），另一部分由 GO/PPY 水凝胶片组成可以发生响应性变形（B）。在红外光的刺激下，软步行机器人的 B 部分发生弯曲，A 部分与水平面紧密接触，当红外光移开时，B 部分会恢复到笔直状态，从而与水平面摩擦，带动整个身体向前。这款软步行机器人展现出通过外部光控实现向前攀爬，并调节持续刺激的时间实现长距离爬行过程。以上结果表明，通过实用的设计和对周围环境的精确控制，开发智能步行设备是完全可行的。

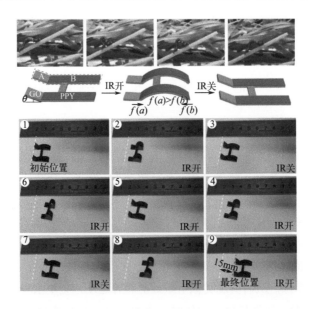

图 9.45　光响应性 GO/PPY 水凝胶软体行走机器人运动的示意图和图像[135]

9.3.1.3 变形器或形状记忆装置

自然界存在许多生物，可以根据其内部的各向异性结构，对外界的攻击或环境的变化做出反应，并产生复杂的形态变化或位置移动。受此启发，研究人员设计了具有多种智能驱动性能的仿生水凝胶执行器[140,141]。Zhao 等[142] 设计和组装了三层水凝胶执行器，并完全模仿菊花花瓣的形状和排列方式，其中菊花的雌蕊和茎分别由真实花的部分和聚二甲基硅氧烷棒组成，将这些组件组装成具有完整结构的光响应仿生水凝胶"花"（图 9.46）。在外部红外光刺激下，光刺激的响应层的"花瓣"会聚集在"花蕊"周围，向"花蕊"方向弯曲变形，随着照射时间的增加以及红外光斑逐渐从内层移到外层，整个"花朵"被近红外光斑覆盖，"花瓣"紧紧地排列在"花蕊"周围，表现为完全闭合的状态。移除光斑后，封闭的菊花逐渐伸展并恢复到原来的形态。整个过程模拟了花朵的完整开放过程，为以后研究实体花朵的开放和变形过程提供了参考和研究思路。

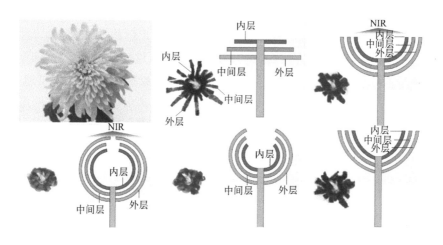

图 9.46　近红外驱动的仿生水凝胶菊花的示意图和照片[142]

Ma 等[143] 受变色龙、章鱼以及金银花变形和开放过程的启发，报道了一种协调变形和变色的宏观各向异性双水凝胶执行器，这种水凝胶执行器是由超支化聚乙烯亚胺（PBI-HPEI）荧光水凝胶层结合超分子主客体聚合物通过热响应性氧化物制备而得，具有 pH 响应性。除了可以实现简单的形状变形外，还可以基于同样的策略构建具有复杂形状转换的执行器。水凝胶执行器层被外层响应水凝胶封闭，在添加外部光刺激时展现出响应变形行为，从而舒展形状，实现类似开花的过程，同时还可以显示出 pH 响应荧光颜色变化（图 9.47）。因此该水凝胶执行器可以模拟一些可以移动的自然生物，展现出有趣的研究应用方向。

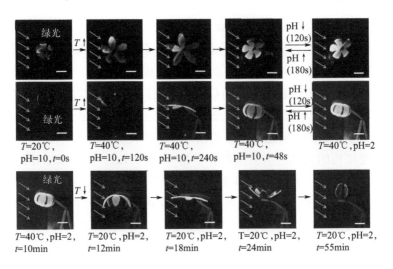

图 9.47　可调节荧光各向异性水凝胶执行器作为形状变换器的照片（比例尺为 5mm）[143]

类似地，通过使用聚丙烯酰胺／聚丙烯酸（PAM/PAAc）双层水凝胶执行器，Hua 等[95]制备了一种不对称的双层水凝胶执行器，顶层是双网络结构，底层是单网络结构。由于紫外线穿透性差，AAc 的光聚合只进行到一定的照射深度，从而形成双层结构，而 AM 形成单网络层。这种水凝胶执行器表现出 UCST 类型的体积相变行为，因此双层 PAM/PAAc 水凝胶的介质可以在其两层之间进行交换，从而实现不受环境约束的应用。当水凝胶的顶部受到紫外线的刺激时，无论在水或液状石蜡的介质中，这种水凝胶"花"在 60℃时可以打开，在 5℃时关闭（图 9.48），再次加热到 60℃后则会发生与闭合相反的过程。然而，水凝胶在不同介质中表现出不同的反应时间差，在水浴中大约需要 10min，在油浴中大约需要 50min。因此，这种不对称水凝胶执行器可以实现两栖驱动性能，这将进一步拓宽该类型水凝胶执行器的潜在应用。

图 9.48　具有自主作用行为的多响应型水凝胶花朵的照片[95]

为了制备出更复杂的形状变形水凝胶，Sun 等[144]通过 3D 打印制备了磁性水凝胶软结构。将硬磁填料 NDFeB 嵌入水凝胶基质并添加纳米胶体改变前驱体的流变性，通过 3D 打印使水凝胶得到各种复杂形状。最有观赏性和代表性的是打印得到的水凝胶"莲花"；随着外部磁场的变化，莲花形状的水凝胶"花朵"会不断地变形改变整体的形状（图 9.49）。在没有磁场的情况下，3D 打印的磁性水凝胶"莲花"是一个完全绽放的状态，随着施加磁场方向从下向上变化，水凝胶"莲花"同步呈现打开到闭合的改变，这种变化最快可达 0.1s，展示出了很高的响应效率。通过不同的外场控制和形状设计的水凝胶执行器结合组装和设计实现复杂的变化过程，为复杂爬行器和形状记忆装置的制备提供了新方法。

图 9.49　磁性水凝胶莲花的致动示例（比例尺为 5mm）[144]

9.3.1.4　折纸设备和可穿戴设备

折纸结构结合了平面制造方法的简单性和折叠可以实现的几何复杂性。水凝胶折纸是一种有趣的尝试。Zhao 等[142]制备了一种用氧化石墨烯（GO）颗粒作光热传感器并可响应 NIR 驱动的水凝胶材料，并设计了一个在弯折处黏附水凝胶片的纸箱（图 9.50）。纸箱包含六部分，每部分的折纸角度可以达到等于或大于 90°的铰链角度，以确保完全折叠状态。在 NIR 照射下，设计的纸箱依次折叠每个部分，将具有响应性的水凝胶执行器应用于执行器的外围以执行变形结果，为确保连续性和准确性，使用三个近红外光点照射水凝胶盒的相应部分，在近红外光的刺激下，展开的纸箱按设计顺序逐渐折叠。去掉外界刺激 15s 后，折叠纸盒即可恢复到展开状态，表现出可控和可重复的特性。这为远程控制复杂行为和执行过程提供了新思路。

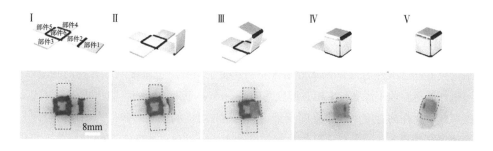

图 9.50　水凝胶执行器作为折纸装置的机理和性能[142]

在自然界中，芫荽草的叶子可以在 1s 内快速闭合以捕捉昆虫，这不同于传统水凝胶相对缓慢的形状转变过程。为了模仿这种快速反应行为，Fan 等[145]使用紫外线诱导 GO 自由基生成并引发聚 N, N- 二甲基氨基乙基甲基丙烯酸酯（PDMAEMA）和 N, N'- 亚甲基双丙烯酰胺（交联剂）聚合形成梯度 GO/PDMAEMA 复合水凝胶执行器。由于 GO 的光热效应，复合水凝胶对 NIR 光有响应行为。当 NIR 照射时，由于最初的双梯度效应，水凝胶片明显地向一侧弯曲，在水下这种复合水凝胶执行器也展现出响应行为，通过不同时间的浸泡可以呈现出各种复杂的折叠形式（图 9.51）。基于这种重复可塑性，他们又设计了复杂的折叠和可编程立方体折叠结构，通过外界刺激，这种设计好的结构可以实现复杂的折纸行为。使用增材制造来制备更复杂的折纸功能和精细结构调节，并将这种水凝胶执行器用于生物医学工具、软机器人、自适应光学和能源系统等。

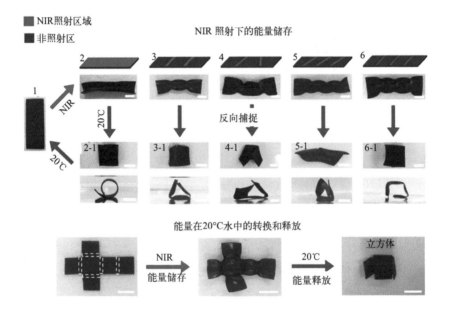

图 9.51　作为复杂形状折纸装置的双梯度复合水凝胶的示意图和照片（比例尺为 1cm）[145]

皮肤界面材料需要低刚度和易变形，这使得水凝胶执行器非常适合作为可穿戴设备等典型的皮肤界面材料。Wang 等[146]使用最常见的大肠杆菌细胞制作水凝胶生物膜并结合天然乳液作为水分惰性材料制备了具有出色湿度响应行为的制动器。当执行器薄膜在干燥条件 [相对湿度（RH），15%] 下，由于细胞脱水收缩，导致薄膜弯曲，而在高湿度下薄膜立即变平。利用这种可逆的重复弯曲行为，他们设计了一种可穿戴设备——带有透气性执行的运动服，可以通过改变织物的温度来改变开关状态用来调节运动服的舒适性（图 9.52），进一步通过形状变化来调整暴露皮肤的面积，以达到皮肤暴露的百分比，从而调节人体的环境舒适程度。在机体高温出汗时，这种智能衣物可以大面积通风加强空气对流，快速除湿；高开孔率保证了动态织物响应的透气性，可以巧妙地反映和响应物理状态，增强对基体核心温度的控制。当测试者开始感到潮湿时，5min后织物的通风瓣打开从而有效去除身上的汗水，降低身体与面料之间的温度。根据相似的思路，Rivera等[147]设计了一种在脱水过程中收缩的水凝胶-纺织品双层执行器，用作天气触发的自适应舒适性的织物服装，用于复杂环境下自动调节功能，使得人体处于舒适状态。

图9.52 在服装原型上使用的汗液响应通风瓣执行器的示意图和图像[146]

9.3.2 流体操控

9.3.2.1 智能阀门

由特定外界刺激引起的形状变化行为意味着水凝胶执行器是流体装置中阀门的出色选择。最直接的实现方式是直接将水凝胶块作为智能阀的控制器。Yu等[148]提出了一种仿生水凝胶阀（图9.53），用于触发和控制微流体通道中流体的流动和停止，并且可以使流体pH、流体流动方向和状态得到精确控制。利用光聚合制备了对pH敏感的水凝胶块，这种水凝胶块在高pH时膨胀，在低pH时收缩，在pH值约5时发生体积位移，从而实现流体流动状态的调节。当从后面施加压力时，阀门处于关闭状态，限制回流，而向前的压力可以使阀门叶片打开并允许流体通过。因此，该阀门可用作单向平衡阀并提供在特定条件下操作的能力。此功能类似于哺乳动物静脉中的止回阀。

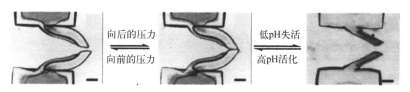

图9.53 水凝胶执行器用于流体控制阀门的图像（比例尺为500μm）[148]

为了实现更复杂的阀门控制系统，Cheng 等[149]报道了一种双层结构的"窗帘式"和"花瓣式"水凝胶瓣膜，在近红外辐射下可分为多个条纹并呈现"开口"。这种双层水凝胶盘固定在通道中，在照射下产生局部变形。"窗帘式"水凝胶在管道中有四个分段（图 9.54），NIR 光的刺激导致 PNIPAM 层局部收缩四个分段依次发生弯曲变形，进而打开通道。通过调节光照强度和照射面积，可以控制折叠打开部分的数量为两个（1.75W）、三个（2.50W）和四个（2.76W），从而可以选择不同的能量来调控水凝胶变形和通道打开状态。其中，单臂实现水平到垂直的弯曲变形约为 90s，而多臂完全变形只需 30s。同样，"花瓣"水凝胶执行器也可以通过光辐照打开和关闭。水凝胶圆盘被切成六部分，初始状态为闭合，形成"花瓣"水凝胶，在近红外光刺激下"花瓣"发生部分变形，从而打开部分阀门。这种水凝胶阀门具有良好的阻隔能力，可用作固体/液体输送和受控响应开关。在进一步的研究中，将 MnO_2 悬浮液添加到水凝胶上方的通道中。通过 NIR（2.76W）照射控制水凝胶阀门打开，并控制 MnO_2 颗粒的通过和节流。这种可控颗粒输送和阀门开关的功能启发了更多的阀门相关的制备。

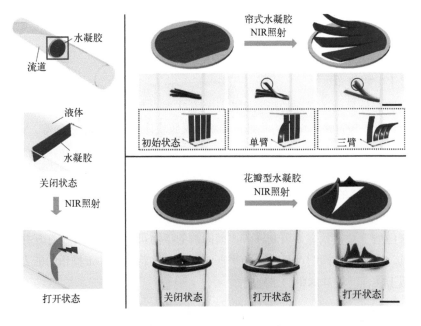

图 9.54 自修复石墨烯氧化物基水凝胶作为光驱动阀门的示意图和照片（比例尺为 1cm）[149]

基于以上响应阀门思路，Wang 等[150]制备了一种带有水凝胶止回阀的新型生物驱动微型泵（图9.55）。这种微型泵为T形结构，包括入口通道、出口通道以及两个通道间的驱动通道，并准备了两个水凝胶球作为止回阀。该泵的工作原理类似于活塞泵，水凝胶活塞被困在专门设计的微通道内，它可以在通道两端之间来回移动，当活塞在通道中向后移动时，流体通过入口阀进入，整个通道处于通路状态，在活塞的往复运动下，流体连续从入口流向出口。在荧光标记微球的具体流动过程中，从入口通道跟踪到出口通道，止回阀对活塞的运动做出快速响应，并在通道中形成定向流体流动。

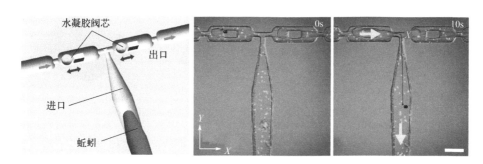

图9.55 作为简单微型止回阀使用的光图案水凝胶的
示意图和照片（比例尺为200μm）[150]

9.3.2.2 微流控

微流控是指微小尺度下控制和操纵流体，广泛应用于生物分子和细胞分析、高通量筛选、诊断和治疗等。一般情况下，微流控设备需要微型智能组件来提供低功率、低成本、不受限制和微型化的定向流体。因此，将柔软和刺激响应的水凝胶集成到微流控装置中是一个极好的思路[151-153]。

响应水凝胶驱动微流控装置中的流体的概念很有吸引力。但是装置可以实现的空间分辨率、速度、可靠性和集成度等限制阻碍了它在过去20年的发展。基于此，D'Eramo等[154]在平面基板上成功制备了PNIPAM水凝胶薄膜并图案化，使用该技术获得了具有多达7800个隔间和释放溶质以及单独驱动的设备，其中每个通道的关闭和打开切换时间分别为（0.6±0.1）s和（0.25±0.15）s。水凝胶片长度超过通道宽度，通过将矩形水凝胶片集成到垂直于其长度的微通道中来充当密封微流体阀门（图9.56）。通过对底板上的材料施加低功率刺激，通道随即关闭，从而防止泄漏。图9.56显示了阀门在打开和关闭位置的荧光显微镜图像和相应的形态。当体系温度在40℃时，高于LCST，水凝胶阀门打开，当温度降至25℃时，水凝胶膜膨胀变大，阀门关闭。该技术可应用于单细胞治疗和基因领域的核酸扩增检测。

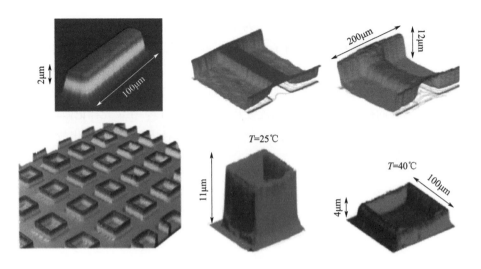

图 9.56 作为微流控执行器使用的温度响应型水凝胶的机理[154]

更进一步地，Takeuchi 等[155] 提出了一种微纤维状的可编程材料用于微流控装置（图 9.57）。这种装置包含用于预凝胶溶液的微通道、用于交联剂的微通道以及一个带有薄膜的气动阀（通过另外两个微通道施加气压控制开关状态）。微纤维上的阀门的开关可以通过计算机控制的编码序列来设计。将两种类型的预凝胶溶液加入装置中并通过第一个汇合点，经图案化微通道后分流为多个预凝胶溶液流，在第二合并点与交联剂 $CaCl_2$ 溶液结合以固化预聚合溶液，最终获得轴向图案化水凝胶微纤维。这为实现复杂的水凝胶执行器控制的反应器和流体精确控制提供了可能。

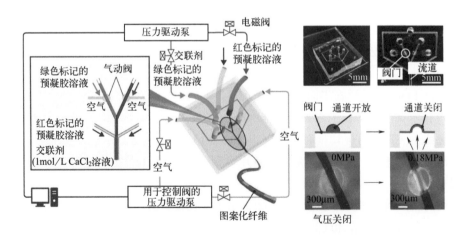

图 9.57 作为阀控微流控系统的水凝胶微纤维的示意图[155]

9.3.3 医学工程

9.3.3.1 定位导航

生物相容性的水凝胶材料在组织工程中是至关重要的组成部分,研究人员已经开发了广泛的水凝胶导航定位器以用于医药组织工程。例如,Tang 等[156] 设计了一种由磁性纳米颗粒和 PNIPAM 热高分子网络组成的具有灵敏响应的水凝胶执行器(图9.58)。这种磁性水凝胶在密封装置中时,磁场可以定向驱动其运动,并且水凝胶可以在环境温度变化中发生复杂的变形行为,因此可以实现更多的应用场景。通过应用直接磁场和交变磁场实现了水凝胶执行器的导航和变形功能,将一个星形制动器放置在一个装满水的弯曲玻璃管中,使用磁铁将执行器从管子引导到被线圈包围的另一端,到线圈中心后,施加交变磁场,只需几分钟,执行器就可以向上弯曲变形从而释放物体。这为水凝胶执行器在管路导航和运动递送中的应用提供了一些思路。

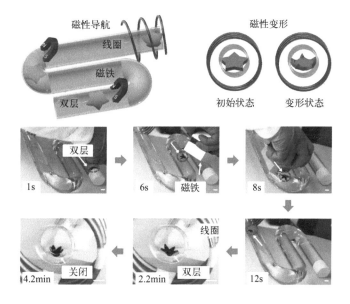

图 9.58　用于磁导航定位的磁热敏水凝胶的机理和性能(比例尺为 10mm)[156]

封闭空间作业(通常是封闭区域,通道受限)在工业应用中十分常见,但有限空间对软机器人的运动构成了相当大的挑战,特别是当所需的机器人大于通道的大小时。为了解决这个问题,Le 等[122] 将 Fe^{3+} 引入高分子中,作为羧基交联的动态骨架,得到透明柔软的高分子,并将 NDFeB 添加到预凝胶中形成均匀分散体,通过进一步聚合形成具有优异响应性能的水凝胶执行器。该执行器可以实现分解导航组件解决尺寸限制的问题。也就是说,大块执行器可以分解成几个部分以有效地进入封闭空间，这些部件都具有非接触式磁性可控的性能，如图 9.59 所示，在封闭的空间内设置了狭窄通道，用于防止水凝胶软机器人通过通道将货物运送到目的地。软机器人首先被磁化并切成棒状的几部分，然后用磁场控制每个部分分别进入"S"管通道，基于磁性组装和快速自愈的特性，多个部分组装焊接成多腿机器人，这个多腿机器人可以将货物运送到目的地。这种执行器展现出复杂组装和任务递送的功能，并且很好地突破了通道空间尺寸的限制，为未来这类执行器开辟了新道路。

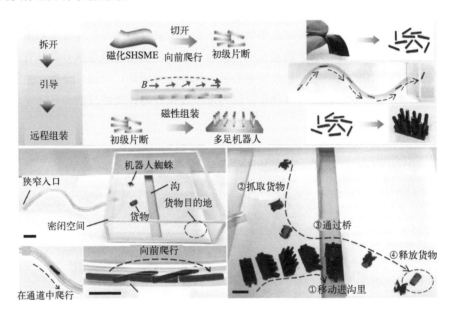

图 9.59 自修复磁执行器用作分离 - 导航 - 组装策略的示意图和照片（比例尺为 10mm）[122]

9.3.3.2 药物输送

长期以来，水凝胶执行器因其对周围环境的灵敏响应性能而被广泛用作药物载体。然而，实际的药物递送过程需要以可编程方式控制药物释放的时间和数量，这是传统响应性水凝胶所缺乏的功能。基于此，Sun 等[144]展示了一种基于 NDFeB 磁性颗粒的增强聚丙烯酰胺（PAM）水凝胶，这种水凝胶可以用作 3D 打印墨水来实现复杂结构的印刷。这种水凝胶执行器可以使用模板辅助磁化策略对微型机器人的磁场进行编程，编程后的微型机器人在外磁场作用下显示精确可控的变形。该六臂水凝胶机器人在生物医学和猪器官异物清除实验中具有潜在应用，应用的过程分为四个步骤：放置机器人、磁导航运动、抓取并搬运物体以及释放物体（图 9.60）。使用卷曲的金属丝作为异物，手持内窥镜用于捕获内部图像并指导软体机器人的操作。将微型机器人放置在器官的开口中，在外部磁场的引导下向器官内部移动，在到达指定位置后通过控制温度实现水凝胶机器人变形并捕获异物，随后将异物移动到所需的位置。该操作可以重复进行，并且水凝胶机器人能够顺利通过褶皱的通道，此外在装满水的器官内充分证明了异物去除能力。尽管磁性水凝胶机器人已经显示出在不同器官中的实际应用，但稳定性仍有待进一步研究。另外考虑到 NDFeB 颗粒的毒性，可以对颗粒施加功能涂层，将更多功能化的生物相容性水凝胶与增强 3D 打印相结合，对于制造更复杂的受控机器人是有希望的。

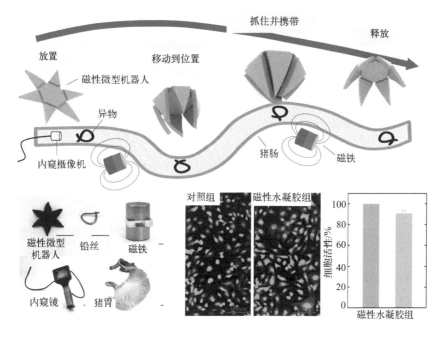

图 9.60　用于异物清除的 3D 打印磁性六臂水凝胶机器人的示意图和图像[144]

Kobayashi 等[157]制备了热响应光学交联高溶胀聚［低聚乙二醇甲基醚甲基丙烯酸酯-双（2-甲基丙烯酰基）氧乙基二硫化物］［P（OEGMA-DSDMA）］凝胶层和掺杂有 Fe_2O_3 纳米颗粒的低膨胀聚［丙烯酰胺 N, N'-双（酰基）胱胺］［P（AM-BAC）］凝胶层的双层水凝胶执行器，其中 P（OEGMA-DSDMA）凝胶层具有 LCST 响应性能可在 20～90℃ 调节。利用该双层水凝胶执行器制备了星形水凝胶夹具（图 9.61），该夹具最初呈现折叠关闭状态，在加热时逐渐打开，当温度冷却下来时重新折叠关闭从而实现夹取物体的功能，这种夹具除了实现拾取和放置任务外，还可以用于药物输送。此外，多指执行器还可以抓取药物，并起到缓释药物的作用。以上这些水凝胶执行器为复杂管道或体内的药物递送和异物清除等应用提供了新的解决方案，但是由于材料选择的限制，还会存在一些问题需要未来的研究人员解决。

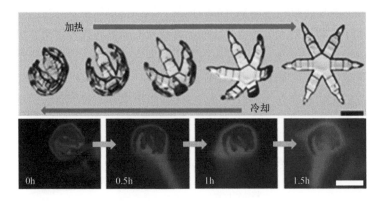

图 9.61　用于药物传递的热磁响应型水凝胶夹持器的示意图和照片（比例尺为 2mm）[157]

9.3.3.3　细胞操控

针对微粒操控和细胞捕获应用需求[158]，Li 等[119] 开发了一种由可弯曲的水凝胶微管组成的 pH 刺激响应执行器（图 9.62）。该水凝胶执行器由四个水凝胶微管组成，其状态可以由伸直（打开状态）和弯曲到同一点（闭合状态）不断切换，从而实现捕获和释放目标物体的功能，其中每个微管的直径为 10μm，高度为 60μm，这些参数可以根据受控粒子的大小进行调整。为了证明该执行器的粒子捕获能力，通过将含直径 13μm PS 微球的溶液滴入开放的微结构中，当加入稀酸溶液（pH＜7）时可激发微管变形，实现捕获的功能，防止颗粒被冲走；当加入稀碱溶液（pH＞9）时使夹子展开，颗粒释放出来。更重要的是，该装置还可以抓取单个细胞，具体操作如下：执行器最初处于打开状态以防止细胞死亡，在

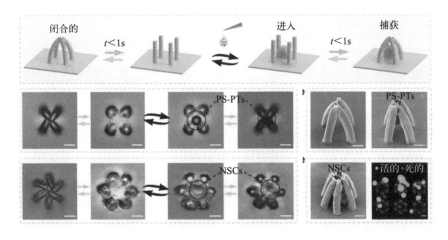

图 9.62　pH 响应性水凝胶微管用于捕获细胞的图像（比例尺为 10μm）[119]

细胞下沉进入捕获之前，向细胞中加入稀盐酸（pH < 7），并关闭捕获器以捕获细胞，第二天测量活/死细胞以确定细胞状态；如图 9.62 所示，绿色、红色（实心）和红色（空心）分别代表活细胞、死细胞和水凝胶微观结构，最终的数据显示约 75% 的细胞存活率，说明本次 pH 触发捕获实验可以用于捕获细胞。因此，这种微型水凝胶执行器提供了一种新的细胞和颗粒操控方法，但也存在需要较大范围的 pH 调控和一定细胞毒性的限制。

然而，在控制细胞过程中，操纵器的细胞毒性和降解需要重点考虑。基于此，Zhao 等[159]报道了一种可生物降解的由形状记忆高分子聚（L-丙交酯-co-D, L-丙交酯）（PLLADLLA）的形状记忆层和包含金纳米棒的光热层组成的双层凝胶微阵列执行器。为了触发形态变化以执行特定相位的细胞操作，利用 NIR 照射金纳米棒层受热从而驱动 PLLADLLA 层发生相变，上部形状记忆层从弯曲阵列形态返回到竖直微观状态，从而实现对于细胞生长的控制和影响（图 9.63）。无 NIR 照射时，来自微型执行器阵列的细胞骨架组织和细胞成核不受限制；在 10s NIR 照射后，细胞骨架组织的成核表现出排列的显著变化，已由各向异性向同一方向改变。因此，在生理温度（37℃）下，动态平台近红外控制的形貌转换可以触发人体细胞的非自发几何回缩。以上微型阵列水凝胶执行器平台不仅可以促进新一代细胞操作平台在各种生物医学领域中的应用，也可以在伤口愈合和各种组织再生过程中起到相应的作用。

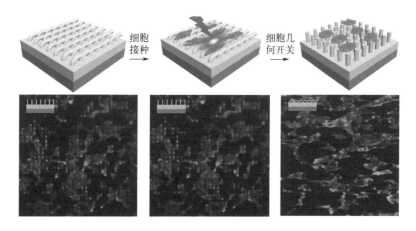

图 9.63 用于细胞几何形状切换的微结构水凝胶平台[159]

参考文献

[1] Lee H, Um D S, Lee Y, et al. Octopus-inspired smart adhesive pads for transfer printing of semiconducting nanomembranes. Advanced Materials, 2016, 28 (34): 7457-7465.

[2] Spinks G M. Advanced actuator materials powered by biomimetic helical fiber topologies. Advanced Materials, 2020, 32 (18): 1904093.

[3] Zhu Q L, Du C, Dai Y, et al. Light-steered locomotion of muscle-like hydrogel by self-coordinated shape change and friction modulation. Nat Commun, 2020, 11 (1): 1-11.

[4] Ren Z, Hu W, Dong X, et al. Multi-functional soft-bodied jellyfish-like swimming. Nat Commun, 2019, 10 (1): 1-12.

[5] Zong L, Li X, Han X, et al. Activation of actuating hydrogels with WS2 nanosheets for biomimetic cellular structures and steerable prompt deformation. ACS Appl Mater Interfaces, 2017, 9 (37): 32280-32289.

[6] Holmes D P, Crosby A J, Snapping surfaces. Advanced Materials, 2007, 19 (21): 3589-3593.

[7] Ionov L, Biomimetic hydrogel-based actuating systems. Advanced Functional Materials, 2013, 23 (36): 4555-4570.

[8] Ma Y, Ma S, Yang W, et al. Sundew-Inspired Simultaneous Actuation and Adhesion/Friction Control for Reversibly Capturing Objects Underwater. Advanced Materials Technologies, 2019, 4 (2): 1800467.

[9] Le X, Lu W, Zhang J, et al. Recent progress in biomimetic anisotropic hydrogel actuators. Advanced science, 2019, 6 (5): 1801584.

[10] Wong W S, Li M, Nisbet D R, et al. Mimosa Origami: A nanostructure-enabled directional self-organization regime of materials. Science advances, 2016, 2 (6): e1600417.

[11] Xu Q, Su R, Chen Y, et al. Metal charge transfer doped carbon dots with reversibly switchable, ultra-high quantum yield photoluminescence. ACS Applied Nano Materials, 2018, 1 (4): 1886-1893.

[12] Ikejiri S, Takashima Y, Osaki M, et al. Solvent-free photoresponsive artificial muscles rapidly driven by molecular machines. Journal of the American Chemical Society, 2018, 140 (49): 17308-17315.

[13] Lin H, Ma S, Yu B, et al. Fabrication of asymmetric tubular hydrogels through polymerization-assisted welding for thermal flow actuated artificial muscles. Chemistry of Materials, 2019, 31 (12): 4469-4478.

[14] Tian Y, Wei X, Wang Z J, et al. A facile approach to prepare tough and responsive ultrathin physical hydrogel films as artificial muscles. ACS Appl Mater Interfaces, 2017, 9 (39): 34349-34355.

[15] Guo M, Wu Y, Xue S, et al. A highly stretchable, ultra-tough, remarkably tolerant, and robust self-healing glycerol-hydrogel for a dual-responsive soft actuator. Journal of Materials Chemistry A, 2019, 7 (45): 25969-25977.

[16] Qin H, Zhang T, Li N, et al. Anisotropic and self-healing hydrogels with multi-responsive actuating capability. Nat Commun, 2019, 10 (1): 1-11.

[17] Wei S, Lu W, Le X, et al. Bioinspired synergistic fluorescence-color-switchable polymeric hydrogel actuators. Angewandte Chemie, 2019, 131 (45): 16389-16397.

[18] Li C, Li M, Ni Z, et al. Stimuli-responsive surfaces for switchable wettability and adhesion. Journal of the Royal Society Interface, 2021, 18 (179): 20210162.

[19] Duan X, Yu J, Zhu Y, et al. Large-scale spinning approach to engineering knittable hydrogel fiber for soft robots. ACS Nano, 2020, 14 (11): 14929-14938.

[20] Cheng Y, Chan K H, Wang X Q, et al. Direct-ink-write 3D printing of hydrogels into biomimetic soft

robots. ACS Nano, 2019, 13 (11): 13176-13184.

[21] Wang X, Jiao N, Tung S, et al. Photoresponsive graphene composite bilayer actuator for soft robots. ACS Appl Mater Interfaces, 2019, 11 (33): 30290-30299.

[22] Yuk H, Lin S, Ma C, et al. Hydraulic hydrogel actuators and robots optically and sonically camouflaged in water. Nature communications, 2017, 8 (1): 1-12.

[23] Chen G, Huang J, Gu J, et al. Highly tough supramolecular double network hydrogel electrolytes for an artificial flexible and low-temperature tolerant sensor. J Mater Chem A, 2020, 8 (14): 6776-6784.

[24] Liang Y, Ye L, Sun X, et al. Tough and stretchable dual ionically cross-linked hydrogel with high conductivity and fast recovery property for high-performance flexible sensors ACS Appl Mater Interfaces, 2019, 12 (1): 1577-1587.

[25] Wang C, Zhang P, Xiao W, et al. Visible-light-assisted multimechanism design for one-step engineering tough hydrogels in seconds. Nat Commun, 2020, 11 (1): 1-9.

[26] Xu L, Wang C, Cui Y, et al. Conjoined-network rendered stiff and tough hydrogels from biogenic molecules. Science advances, 2019, 5 (2): eaau3442.

[27] Zhang H J, Sun T L, Zhang A K, et al. Tough physical double-network hydrogels based on amphiphilic triblock copolymers. Advanced Materials, 2016, 28 (24): 4884-4890.

[28] Lei Z, Wang Q, Wu P. A multifunctional skin-like sensor based on a 3D printed thermo-responsive hydrogel. Materials Horizons, 2017, 4 (4): 694-700.

[29] Matsumoto K, Sakikawa N, Miyata T. Thermo-responsive gels that absorb moisture and ooze water. Nat Commun, 2018, 9 (1): 1-7.

[30] Zhang X, Soh S. Performing Logical Operations with Stimuli-Responsive Building Blocks. Advanced Materials, 2017, 29 (18): 1606483.

[31] Yesilyurt V, Webber M J, Appel E A, et al. Injectable self-healing glucose-responsive hydrogels with pH-regulated mechanical properties. Advanced materials, 2016, 28 (1): 86-91.

[32] Kahn J S, Hu Y, Willner I. Stimuli-responsive DNA-based hydrogels: from basic principles to applications. Acc Chem Res, 2017, 50 (4): 680-690.

[33] Li L, Scheiger J M, Levkin P A. Design and applications of photoresponsive hydrogels. Advanced Materials, 2019, 31 (26): 1807333.

[34] Jiang Z, Tan M L, Taheri M, et al. Strong, Self-Healable, and Recyclable Visible-Light-Responsive Hydrogel Actuators. Angewandte Chemie, 2020, 132 (18): 7115-7122.

[35] Tao N, Zhang D, Li X, et al. Near-infrared light-responsive hydrogels via peroxide-decorated MXene-initiated polymerization. Chem Sci, 2019, 10 (46): 10765-10771.

[36] Zhang H, Zeng H, Priimagi A, et al. Programmable responsive hydrogels inspired by classical conditioning algorithm. Nat Commun, 2019, 10 (1): 1-8.

[37] Downs F G, Lunn D J, Booth M J, et al. Multi-responsive hydrogel structures from patterned droplet networks. Nat Chem, 2020, 12 (4): 363-371.

[38] Manjua A C, Alves V D, Crespo J O G, et al. Magnetic responsive PVA hydrogels for remote modulation of protein sorption. ACS Appl Mater Interfaces, 2019, 11 (23): 21239-21249.

[39] Li Y, Sun Y, Xiao Y, et al. Electric field actuation of tough electroactive hydrogels cross-linked by functional triblock copolymer micelles. ACS Appl Mater Interfaces, 2016, 8 (39): 26326-26331.

[40] Qu J, Zhao X, Ma P X, et al. Injectable antibacterial conductive hydrogels with dual response to an electric field and pH for localized "smart" drug release. Acta biomaterialia, 2018, 72: 55-69.

[41] Liu H, Rong L, Wang B, et al. Facile fabrication of redox/pH dual stimuli responsive cellulose hydrogel. Carbohydrate polymers, 2017, 176: 299-306.

[42] Qiao L, Wang X, Gao Y, et al. Laccase-mediated formation of mesoporous silica nanoparticle based redox stimuli-responsive hybrid nanogels as a multifunctional nanotheranostic agent. Nanoscale, 2016, 8 (39): 17241-17249.

[43] Shigemitsu H, Fujisaku T, Onogi S, et al. Preparation of supramolecular hydrogel-enzyme hybrids exhibiting biomolecule-responsive gel degradation. Nat Protoc, 2016, 11 (9): 1744-1756.

[44] Li C, Feng S, Li C, et al. Synthesizing Organo/Hydrogel Hybrids with Diverse Programmable Patterns and Ultrafast Self-Actuating Ability via a Site-Specific "In Situ" Transformation Strategy. Advanced Functional Materials, 2020, 30 (32): 2002163.

[45] Li M, Wang X, Dong B, et al. In-air fast response and high speed jumping and rolling of a light-driven hydrogel actuator. Nat. Commun, 2020, 11 (1): 1-10.

[46] Zhang M, Wang Y, Jian M, et al. Spontaneous alignment of graphene oxide in hydrogel during 3D printing for multistimuli-responsive actuation. Adv Sci, 2020, 7 (6): 1903048.

[47] Hines L, Petersen K, Lum G Z, et al. Soft actuators for small-scale robotics. Advanced materials, 2017, 29 (13): 1603483.

[48] Wang B, Facchetti A. Mechanically flexible conductors for stretchable and wearable e-skin and e-textile devices. Advanced Materials, 2019, 31 (28): 1901408.

[49] Ma Y, Lin M, Huang G, et al. 3D spatiotemporal mechanical microenvironment: a hydrogel-based platform for guiding stem cell fate. Advanced Materials, 2018, 30 (49): 1705911.

[50] Forterre Y, Skotheim J M, Dumais J, et al. How the Venus flytrap snaps. Nature, 2005, 433 (7024): 421-425.

[51] Fratzl P, Barth F G. Biomaterial systems for mechanosensing and actuation. Nature, 2009, 462 (7272): 442-448.

[52] Jiang Y, Korpas L M, Raney J R. Bifurcation-based embodied logic and autonomous actuation. Nat Commun, 2019, 10 (1): 1-10.

[53] Wang X, Khara A, Chen C. A soft pneumatic bistable reinforced actuator bioinspired by Venus Flytrap with enhanced grasping capability. Bioinspiration & Biomimetics, 2020, 15 (5): 056017.

[54] Le X, Lu W, Zheng J, et al. Stretchable supramolecular hydrogels with triple shape memory effect. Chem Sci, 2016, 7 (11): 6715-6720.

[55] Volkov A G, Foster J C, Baker K D, et al. Mechanical and electrical anisotropy in Mimosa pudica pulvini. Plant Signaling Behav, 2010, 5 (10): 1211-1221.

[56] Deng J, Li J, Chen P, et al. Tunable photothermal actuators based on a pre-programmed aligned nanostructure. Journal of the American Chemical Society, 2016, 138 (1): 225-230.

[57] Erb R M, Sander J S, Grisch R, et al. Self-shaping composites with programmable bioinspired microstructures. Nat Commun, 2013, 4 (1): 1-8.

[58] Ko H, Javey A. Smart actuators and adhesives for reconfigurable matter. Acc Chem Res, 2017, 50 (4): 691-702.

[59] Taccola S, Greco F, Sinibaldi E, et al. Toward a new generation of electrically controllable hygromorphic soft actuators. Advanced Materials, 2015, 27 (10): 1668-1675.

[60] Dumais J, Forterre Y. "Vegetable dynamicks": the role of water in plant movements. Annual Review of Fluid Mechanics, 2012, 44 (1): 453-478.

[61] Burgert I, Fratzl P. Actuation systems in plants as prototypes for bioinspired devices. Philosophical Transactions of the Royal Society A: Mathematical, Physical and Engineering Sciences, 2009, 367 (1893): 1541-1557.

[62] Elbaum R, Gorb S, Fratzl P. Structures in the cell wall that enable hygroscopic movement of wheat awns. J Struct Biol, 2008, 164 (1): 101-107.

[63] Elbaum R, Zaltzman L, Burgert I, et al. The role of wheat awns in the seed dispersal unit. Science, 2007, 316 (5826): 884-886.

[64] Motokawa T. Effects of ionic environment on viscosity of Triton-extracted catch connective tissue of a sea cucumber body wall. Comp. Biochem. Physiol B: Comp Biochem, 1994, 109 (4): 613-622.

[65] Szulgit G K, Shadwick R E. Dynamic mechanical characterization of a mutable collagenous tissue: response of sea cucumber dermis to cell lysis and dermal extracts. Journal of Experimental Biology, 2000, 203 (10): 1539-1550.

[66] Capadona J R, Shanmuganathan K, Tyler D J, et al. Stimuli-responsive polymer nanocomposites inspired by the sea cucumber dermis. Science, 2008, 319 (5868): 1370-1374.

[67] Trotter J A, Lyons-Levy G, Chino K, et al. Collagen fibril aggregation-inhibitor from sea cucumber dermis. Matrix Biology, 1999, 18 (6): 569-578.

[68] Trotter J A, Lyons-Levy G, Luna D, et al. Stiparin: a glycoprotein from sea cucumber dermis that aggregates collagen fibrils. Matrix biology, 1996, 15 (2): 99-110.

[69] Miyawaki A, Llopis J, Heim R, et al. Fluorescent indicators for Ca^{2+} based on green fluorescent proteins and calmodulin. Nature, 1997, 388 (6645): 882-887.

[70] Shang L, Zhang W, Xu K, et al. Bio-inspired intelligent structural color materials. Materials Horizons, 2019, 6 (5): 945-958.

[71] Zhang Z, Chen Z, Wang Y, et al. Bioinspired conductive cellulose liquid-crystal hydrogels as multifunctional electrical skins. Proceedings of the National Academy of Sciences, 2020, 117 (31): 18310-18316.

[72] Teyssier J, Saenko S V, Van Der Marel D, et al. Photonic crystals cause active colour change in chameleons. Nat Commun, 2015, 6 (1): 1-7.

[73] Zhao P, Chen H, Li B, et al. Stretchable electrochromic devices enabled via shape memory alloy composites (SMAC) for dynamic camouflage. Optical Materials, 2019, 94: 378-386.

[74] Zhang Z L, Dong X, Fan Y N, et al. Chameleon-inspired variable coloration enabled by a highly flexible photonic cellulose film. ACS Appl Mater Interfaces, 2020, 12 (41): 46710-46718.

[75] Montero de Espinosa L, Meesorn W, Moatsou D, et al. Bioinspired polymer systems with stimuli-responsive mechanical properties. Chemical reviews, 2017, 117 (20): 12851-12892.

[76] McHenry M J, Jed J. The ontogenetic scaling of hydrodynamics and swimming performance in jellyfish (Aurelia aurita). Journal of experimental biology, 2003, 206 (22): 4125-4137.

[77] Nawroth J, Feitl K, Colin S, et al. Phenotypic plasticity in juvenile jellyfish medusae facilitates

effective animal–fluid interaction. Biol Lett, 2010, 6 (3): 389-393.

[78] Nawroth J C, Lee H, Feinberg A W, et al. A tissue-engineered jellyfish with biomimetic propulsion. Nature biotechnology, 2012, 30 (8): 792-797.

[79] Gu S, Guo S. Performance evaluation of a novel propulsion system for the spherical underwater robot (SURIII). Applied Sciences, 2017, 7 (11): 1196.

[80] Mather J A. How do octopuses use their arms? J Comp Psychol, 1998, 112 (3): 306.

[81] Sfakiotakis M, Kazakidi A, Tsakiris D. Octopus-inspired multi-arm robotic swimming. Bioinspiration & biomimetics, 2015, 10 (3): 035005.

[82] Cianchetti M, Arienti A, Follador M, et al. Design concept and validation of a robotic arm inspired by the octopus. Materials Science and Engineering: C, 2011, 31 (6): 1230-1239.

[83] Renda F, Cianchetti M, Giorelli M, et al. A 3D steady-state model of a tendon-driven continuum soft manipulator inspired by the octopus arm. Bioinspiration & biomimetics, 2012, 7 (2): 025006.

[84] Guglielmino E, Tsagarakis N, Caldwell D G. In An octopus anatomy-inspired robotic arm, 2010 IEEE/RSJ International Conference on Intelligent Robots and Systems, IEEE, 2010: 3091-3096.

[85] Gong C, Shi S, Dong P, et al. Synthesis and characterization of PEG-PCL-PEG thermosensitive hydrogel. International journal of pharmaceutics, 2009, 365 (1-2): 89-99.

[86] Shi Q, Liu H, Tang D, et al. Bioactuators based on stimulus-responsive hydrogels and their emerging biomedical applications. NPG Asia Mater, 2019, 11 (1): 1-21.

[87] Yao C, Liu Z, Yang C, et al. Poly (N-isopropylacrylamide) -clay nanocomposite hydrogels with responsive bending property as temperature-controlled manipulators. Advanced Functional Materials, 2015, 25 (20): 2980-2991.

[88] Liu F, Urban M W. Recent advances and challenges in designing stimuli-responsive polymers. Progress in polymer science, 2010, 35 (1-2): 3-23.

[89] Schild H G. Poly (N-isopropylacrylamide): experiment, theory and application. Progress in polymer science, 1992, 17 (2): 163-249.

[90] Taylor M J, Tomlins P, Sahota T S. Thermoresponsive gels. Gels, 2017, 3 (1): 4.

[91] Tang L, Wang L, Yang X, et al. Poly (N-isopropylacrylamide) -based smart hydrogels: Design, properties and applications. Progress in Materials Science, 2021, 115: 100702.

[92] He X, Aizenberg M, Kuksenok O, et al. Synthetic homeostatic materials with chemo-mechano-chemical self-regulation. Nature, 2012, 487 (7406): 214-218.

[93] Kim Y S, Liu M, Ishida Y, et al. Thermoresponsive actuation enabled by permittivity switching in an electrostatically anisotropic hydrogel. Nature materials, 2015, 14 (10): 1002-1007.

[94] Nojoomi A, Arslan H, Lee K, et al. Bioinspired 3D structures with programmable morphologies and motions. Nat Commun, 2018, 9 (1): 1-11.

[95] Hua L, Xie M, Jian Y, et al. Multiple-responsive and amphibious hydrogel actuator based on asymmetric UCST-type volume phase transition. ACS applied materials & interfaces, 2019, 11 (46): 43641-43648.

[96] Kloxin A M, Kasko A M, Salinas C N, et al. Photodegradable hydrogels for dynamic tuning of physical and chemical properties. Science, 2009, 324 (5923): 59-63.

[97] Li Q, Fuks G, Moulin E, et al. Macroscopic contraction of a gel induced by the integrated motion of

light-driven molecular motors. Nat Nanotechnol, 2015, 10 (2): 161-165.

[98] Roy D, Brooks W L, Sumerlin B S. New directions in thermoresponsive polymers. Chemical Society Reviews, 2013, 42 (17): 7214-7243.

[99] Li J, Ma Q, Xu Y, et al. Highly bidirectional bendable actuator engineered by LCST-UCST bilayer hydrogel with enhanced interface. ACS Appl Mater Interfaces, 2020, 12 (49): 55290-55298.

[100] Zheng J, Xiao P, Le X, et al. Mimosa inspired bilayer hydrogel actuator functioning in multi-environments. Journal of Materials Chemistry C, 2018, 6 (6): 1320-1327.

[101] Shi Q, Xia H, Li P, et al. Photothermal Surface Plasmon Resonance and Interband Transition-Enhanced Nanocomposite Hydrogel Actuators with Hand-Like Dynamic Manipulation. Advanced Optical Materials, 2017, 5 (22): 1700442.

[102] Yang H, Leow W R, Wang T, et al. 3D printed photoresponsive devices based on shape memory composites. Advanced Materials, 2017, 29 (33): 1701627.

[103] Sun Z, Yamauchi Y, Araoka F, et al. An Anisotropic Hydrogel Actuator Enabling Earthworm-Like Directed Peristaltic Crawling. Angewandte Chemie, 2018, 130 (48): 15998-16002.

[104] Xue P, Bisoyi H K, Chen Y, et al. Near-infrared light-driven shape-morphing of programmable anisotropic hydrogels enabled by MXene nanosheets. Angewandte Chemie International Edition, 2021, 60 (7): 3390-3396.

[105] Shankar A, Safronov A P, Mikhnevich E A, et al. Multidomain iron nanoparticles for the preparation of polyacrylamide ferrogels. Journal of Magnetism and Magnetic Materials, 2017, 431: 134-137.

[106] Kim Y, Yuk H, Zhao R, et al. Printing ferromagnetic domains for untethered fast-transforming soft materials. Nature, 2018, 558 (7709): 274-279.

[107] Roeder L, Bender P, Tschöpe A, et al. Shear modulus determination in model hydrogels by means of elongated magnetic nanoprobes. Journal of Polymer Science Part B: Polymer Physics, 2012, 50 (24): 1772-1781.

[108] Li H, Go G, Ko S Y, et al. Magnetic actuated pH-responsive hydrogel-based soft micro-robot for targeted drug delivery. Smart Materials and Structures, 2016, 25 (2): 027001.

[109] Podstawczyk D, Nizio M, Szymczyk P, et al. 3D printed stimuli-responsive magnetic nanoparticle embedded alginate-methylcellulose hydrogel actuators. Additive Manufacturing, 2020, 34: 101275.

[110] Cheng Y, Chan K H, Wang X Q, et al. A fast autonomous healing magnetic elastomer for instantly recoverable, modularly programmable, and thermorecyclable soft robots. Advanced Functional Materials, 2021, 31 (32): 2101825.

[111] Bassetti M J, Chatterjee A N, Aluru N, et al. Development and modeling of electrically triggered hydrogels for microfluidic applications. Journal of Microelectromechanical Systems, 2005, 14 (5): 1198-1207.

[112] Osada Y, Okuzaki H, Hori H. A polymer gel with electrically driven motility. Nature, 1992, 355 (6357): 242-244.

[113] Yang C, Liu Z, Chen C, et al. Reduced graphene oxide-containing smart hydrogels with excellent electro-response and mechanical properties for soft actuators. ACS applied materials & interfaces, 2017, 9 (18): 15758-15767.

[114] Kang Y W, Woo J, Lee H R, et al. A mechanically enhanced electroactive hydrogel for 3D printing

using a multileg long chain crosslinker. Smart Materials and Structures, 2019, 28 (9): 095016.

[115] Ko J, Kim D, Song Y, et al. Electroosmosis-driven hydrogel actuators using hydrophobic/hydrophilic layer-by-layer assembly-induced crack electrodes. ACS nano, 2020, 14 (9): 11906-11918.

[116] Albright V, Zhuk I, Wang Y, et al. Self-defensive antibiotic-loaded layer-by-layer coatings: Imaging of localized bacterial acidification and pH-triggering of antibiotic release. Acta biomaterialia, 2017, 61: 66-74.

[117] Guo W, Lu C H, Orbach R, et al. pH-Stimulated DNA Hydrogels Exhibiting Shape-Memory Properties. Advanced Materials, 2015, 27 (1): 73-78.

[118] Duan J, Liang X, Zhu K, et al. Bilayer hydrogel actuators with tight interfacial adhesion fully constructed from natural polysaccharides. Soft Matter, 2017, 13 (2): 345-354.

[119] Li R, Jin D, Pan D, et al. Stimuli-responsive actuator fabricated by dynamic asymmetric femtosecond bessel beam for in situ particle and cell manipulation. ACS nano, 2020, 14 (5): 5233-5242.

[120] Xu Y, Bolisetty S, Drechsler M, et al. pH and salt responsive poly (N, N-dimethylaminoethyl methacrylate) cylindrical brushes and their quaternized derivatives. Polymer, 2008, 49 (18): 3957-3964.

[121] Blackman L D, Gunatillake P A, Cass P, et al. An introduction to zwitterionic polymer behavior and applications in solution and at surfaces. Chemical Society Reviews, 2019, 48 (3): 757-770.

[122] Le X X, Lu W, He J, et al. Ionoprinting controlled information storage of fluorescent hydrogel for hierarchical and multi-dimensional decryption. Science China Materials, 2019, 62 (6): 831-839.

[123] Du X, Cui H, Zhao Q, et al. Inside-out 3D reversible ion-triggered shape-morphing hydrogels. Research, 2019.

[124] Dawson C, Vincent J F, Rocca A M.How pine cones open. Nature, 1997, 390 (6661): 668.

[125] Kim H, Lee S J. Stomata-inspired membrane produced through photopolymerization patterning. Advanced Functional Materials, 2015, 25 (28): 4496-4505.

[126] Lv C, Sun X C, Xia H, et al. Humidity-responsive actuation of programmable hydrogel microstructures based on 3D printing. Sensors and Actuators B: Chemical, 2018, 259: 736-744.

[127] Zheng Y, Huang H, Wang Y, et al. Poly (vinyl alcohol) based gradient cross-linked and reprogrammable humidity-responsive actuators. Sensors and Actuators B: Chemical, 2021, 349: 130735.

[128] Wu S, Shi H, Lu W, et al. Aggregation-Induced Emissive Carbon Dots Gels for Octopus-Inspired Shape/Color Synergistically Adjustable Actuators. Angewandte Chemie International Edition, 2021, 60 (40): 21890-21898.

[129] Hubbard A M, Cui W, Huang Y, et al. Hydrogel/elastomer laminates bonded via fabric interphases for stimuli-responsive actuators. Matter, 2019, 1 (3): 674-689.

[130] Ulijn R V, Bibi N, Jayawarna V, et al. Bioresponsive hydrogels. Materials today, 2007, 10 (4): 40-48.

[131] Cangialosi A, Yoon C, Liu J, et al. DNA sequence-directed shape change of photopatterned hydrogels via high-degree swelling. Science, 2017, 357 (6356): 1126-1130.

[132] Kaspar C, Ravoo B, van der Wiel W G, et al. The rise of intelligent matter. Nature, 2021, 594 (7863): 345-355.

[133] Li M, Pal A, Aghakhani A, et al. Soft actuators for real-world applications. Nature Reviews Materials, 2022, 7 (3): 235-249.

[134] Yang C, Suo Z. Hydrogel ionotronics. Nature Reviews Materials, 2018, 3 (6): 125-142.

[135] Dong Y, Wang J, Guo X, et al. Multi-stimuli-responsive programmable biomimetic actuator. Nature communications, 2019, 10 (1): 1-10.

[136] Yao Y, Yin C, Hong S, et al. Lanthanide-ion-coordinated supramolecular hydrogel inks for 3D printed full-color luminescence and opacity-tuning soft actuators. Chemistry of Materials, 2020, 32 (20): 8868-8876.

[137] Zhao H, Huang Y, Lv F, et al. Biomimetic 4D-Printed Breathing Hydrogel Actuators by Nanothylakoid and Thermoresponsive Polymer Networks. Advanced Functional Materials, 2021, 31 (49): 2105544.

[138] Yoon C. Advances in biomimetic stimuli responsive soft grippers. Nano Convergence, 2019, 6 (1): 1-14.

[139] Zhang A, Wang F, Chen L, et al. 3D printing hydrogels for actuators: A review. Chinese Chemical Letters, 2021, 32 (10): 2923-2932.

[140] Armon S, Efrati E, Kupferman R, et al. Geometry and mechanics in the opening of chiral seed pods. Science, 2011, 333 (6050): 1726-1730.

[141] Cheng F M, Chen H X, Li H D. Recent progress on hydrogel actuators. Journal of Materials Chemistry B, 2021, 9 (7): 1762-1780.

[142] Zhao Q, Liang Y, Ren L, et al. Bionic intelligent hydrogel actuators with multimodal deformation and locomotion. Nano Energy, 2018, 51: 621-631.

[143] Ma C, Lu W, Yang X, et al. Bioinspired anisotropic hydrogel actuators with on-off switchable and color-tunable fluorescence behaviors. Advanced Functional Materials, 2018, 28 (7): 1704568.

[144] Sun B, Jia R, Yang H, et al. Magnetic Arthropod Millirobots Fabricated by 3D-Printed Hydrogels. Advanced Intelligent Systems, 2022, 4 (1): 2100139.

[145] Fan W, Shan C, Guo H, et al. Dual-gradient enabled ultrafast biomimetic snapping of hydrogel materials. Science advances, 2019, 5 (4): eaav7174.

[146] Wang W, Yao L, Cheng C Y, et al. Harnessing the hygroscopic and biofluorescent behaviors of genetically tractable microbial cells to design biohybrid wearables. Science Advances, 2017, 3 (5): e1601984.

[147] Rivera M L, Forman J, Hudson S E, et al. In Hydrogel-textile composites: Actuators for shape-changing interfaces, Extended Abstracts of the 2020 CHI Conference on Human Factors in Computing Systems, 2020: 1-9.

[148] Yu Q, Bauer J M, Moore J S, et al. Responsive biomimetic hydrogel valve for microfluidics. Applied Physics Letters, 2001, 78 (17): 2589-2591.

[149] Cheng Y, Ren K, Huang C, et al. Self-healing graphene oxide-based nanocomposite hydrogels serve as near-infrared light-driven valves. Sensors and Actuators B: Chemical, 2019, 298: 126908.

[150] Wang Y, Toyoda K, Uesugi K, et al. A simple micro check valve using a photo-patterned hydrogel valve core. Sensors and Actuators A: Physical, 2020, 304: 111878.

[151] Sackmann E K, Fulton A L, Beebe D J. The present and future role of microfluidics in biomedical

research. Nature, 2014, 507 (7491): 181-189.

[152] Sontheimer-Phelps A, Hassell B A, Ingber D E.Modelling cancer in microfluidic human organs-on-chips. Nature Reviews Cancer, 2019, 19 (2): 65-81.

[153] Ter Schiphorst J, Saez J, Diamond D, et al. Light-responsive polymers for microfluidic applications. Lab on a Chip, 2018, 18 (5): 699-709.

[154] D'Eramo L C, Chollet B, Leman M, et al. Microfluidic actuators based on temperature-responsive hydrogels. Microsystems & Nanoengineering, 2018, 4 (1): 1-7.

[155] Takeuchi N, Nakajima S, Yoshida K, et al. Microfiber-shaped programmable materials with stimuli-responsive hydrogel. Soft Robotics, 2022, 9 (1): 89-97.

[156] Tang J, Yin Q, Qiao Y, et al. Shape morphing of hydrogels in alternating magnetic field. ACS applied materials & interfaces, 2019, 11 (23): 21194-21200.

[157] Kobayashi K, Yoon C, Oh S H, et al. Biodegradable thermomagnetically responsive soft untethered grippers. ACS applied materials & interfaces, 2018, 11 (1): 151-159.

[158] Malachowski K, Jamal M, Jin Q, et al. Self-folding single cell grippers. Nano letters, 2014, 14 (7): 4164-4170.

[159] Zhao Q, Wang J, Wang Y, et al. A stage-specific cell-manipulation platform for inducing endothelialization on demand. National Science Review, 2020, 7 (3): 629-643.

第 10 章

仿生自供能水凝胶传感器

10.1 自供能水凝胶传感器的供能机制
10.2 自供能水凝胶传感器的典型应用
参考文献

水凝胶传感器因其优异的生物相容性和环境适应性而被广泛应用于健康保健、人机交互和环境监测等领域。然而，目前绝大多数的水凝胶传感器主要是依赖于外部能源实现驱动的，这极大地限制了其应用。而基于纳米发电机的能量收集技术的发展为设计和制造可实现能量持续供应的自供电水凝胶传感器铺平了道路。在本章中，我们将探讨基于不同纳米发电效应的自供电传感器的自供电机理及其在水凝胶传感器领域的最新研究进展，重点是最近开发的自供电水凝胶传感器的结构设计、供能机制以及相应的应用场景（图10.1）。

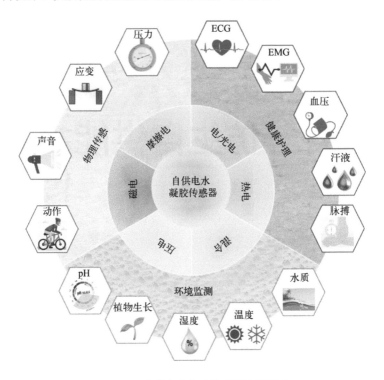

图 10.1　自供电水凝胶传感器的总结图

从发电机制到整个自供电水凝胶传感器系统，从基于单一纳米发电机（NG）效应的水凝胶传感系统到基于混合 NG 效应的水凝胶传感系统，包括物理传感、医疗保健和环境监测等应用场景

10.1　自供能水凝胶传感器的供能机制

深入了解对应的发电机制对于自供电水凝胶电子器件的设计和相应性能的提升至关重要。到目前为止，研究人员已经开发了基于压电、摩擦电、热电、磁

电、光/水伏效应，抑或是结合了其中两种及以上的混合纳米发电效应的各种能量转换技术，并成功地将机械能、热能、磁能、光能或化学能转化为电能。在本节中，我们将集中介绍这些能量收集器的具体工作机制。

10.1.1 摩擦纳米发电

摩擦纳米发电机（TENG）的能量收集结合了两个摩擦层之间的摩擦起电和静电感应现象，因而能有效地将环境中的机械能转化为有价值的电能[1-3]。摩擦层的组成材料通常具有不同的电子吸引能力，当它们接触时，这两个摩擦面就会产生相反电性的电荷。当摩擦层在外力的进一步作用下分离后，因摩擦而产生的电荷会在两个摩擦面之间引起电位降，而且该电位降会随着摩擦电表面的接触和分离而变化。为了平衡上述电位差，自由电子会通过外部电路来回流动，这种交替性的电子流动实现了电流的输出。根据器件结构和工作方式，TENG 可以分为以下四种不同的基本工作模式（图 10.2），分别为横向滑动模式（LS）、垂直接触分离模式（CS）、单电极模式（SE）和独立式摩擦层模式（FT）。其中，LS 和 CS 具有不同的摩擦方向但相同的组成配置，即上层和下层的导电层（电极）以及电介质层（摩擦层）。CS 模式依赖于垂直于界面的两个摩擦电层之间的相对运动，而 LS 模式则依赖于摩擦电层在平行于界面方向上的相对位移。SE 则只有一个电极和一个摩擦层，当外部物体接触到摩擦层时就会发生接触起电。FT 往往是通过滑动电介质层从而与位于其下的相互连接的两个导电层摩擦来实现电流输出。这 4 种不同工作模式的结构特性、优缺点以及应用场景如表 10.1 所示[2]。

表 10.1 TENG 四种工作模式的特点[2]

工作模式	结构特点	优势	劣势	应用场景
LS	水平/旋转运动，几乎没有间隙	高频、连续、高电量输出	摩擦面易损坏，长期稳定性差	旋转，和空气/水流能量收集
CS	垂直运动，存在较大间隙	高输出电压	脉冲输出	挤压、冲击、弯曲、摇晃、振动
SE	只有一个电极，另一个电极接地	简单、多用途的能量收集、易于集成和携带	相对较低的输出和信号的不稳定性	滑动/打字/触屏
FT	多种运动形式，对称电极，不对称电荷分布	能量转化效率高	电极固定不便移动，集成复杂	旋转和振动能量收集

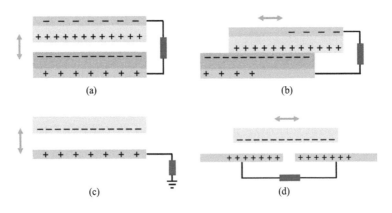

图 10.2　TENG 的四种基本工作模式 [1]

（a）垂直接触分离模式；（b）横向滑动模式；（c）单电极模式；（d）独立式摩擦层模式

10.1.2　压电纳米发电

　　压电纳米发电机（PENG）的工作原理是通过压电材料的压电效应有效地将机械能（应变或应力）转换为电能，从而实现能量采集 [4]。PENG 的基本结构为三层，即上下电极层以及中间的压电材料层（图 10.3）。在外力（直接压电效应）或外加电场（反向压电效应）的作用下，中间层的压电材料会发生机械变形，而这种变形会使得材料内部极化（材料内部的正负电荷中心偏移）并产生体电荷和表面电荷，从而在电极上感应出电位降。该压电势会驱动电子流经外部电路而输出电能 [5]。通过周期性地改变施加外力的大小和方向，感应出的压电势的大小和方向会随之出现周期性变化，这有助于外部电路实现交流电的持续性输出 [6]。压电材料的关键特征是压电系数，压电系数越高则压电材料的能量转换效率越高 [1]。在应变率和器件面积固定的情况下，可以通过采用高压电系数的压电材料和优化柔性基板的设计结构来实现高输出 PENG。

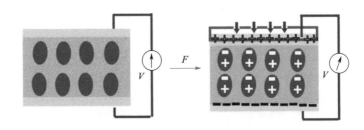

图 10.3　PENG 的工作示意图 [4]

10.1.3 热电纳米发电

热电/热释电纳米发电机（PyNG/TEG）主要是依靠材料的自旋塞贝克效应将热能转化为电能（图 10.4），其核心是导体中载流子的能量随温度的正向变化，温度越高载流子的动能越大[7]。当导体两端接头处的温度存在差异时（ΔT），热端载流子会向冷端移动并在冷端积累，直至载流子的浓度分布达到平衡。此时，导体两端便会产生相应的电动势（ΔV）并在外电路中驱动电子定向移动，从而起到供能的作用。因此，我们可利用人体与环境之间的温差开发出能将人体热量转化为电能的 PyNG，使其作为电源为医疗信息传感器提供电力。

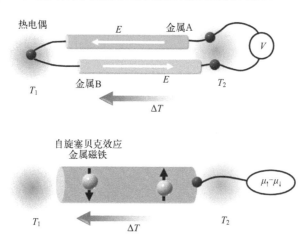

图 10.4 PyNG/TEG 的工作示意图[1]

10.1.4 光伏发电

作为太阳能电池的核心部件，PN 结的光伏效应可以有效地将太阳能转换为电能[8]。如图 10.5 所示，在 PN 结界面，N 型半导体侧存在大量可以扩散到 P 型半导体中的自由电子，而 P 型半导体中则有大量可以扩散到 N 型半导体中的空穴。该扩散过程使得界面处的 N 型半导体带正电，而 P 型半导体带负电，从而在 PN 结界面处形成从 N 型半导体指向 P 型半导体的内置

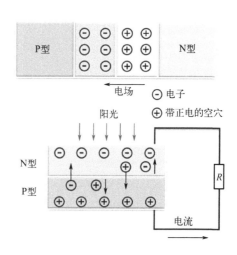

图 10.5 光伏发电的工作示意图[1]

电场。当该器件受到阳光照射时，PN 结界面处会形成电子 - 空穴对。在内置电场的作用下，电子会向 N 型半导体迁移，而空穴则向 P 型半导体迁移，直至 N 型和 P 型半导体分别积累的电子和空穴所产生的光生电场可以抵消内置电场。此时，N 型半导体和 P 型半导体分别带负电和正电，并在 PN 结处形成电位差，使得自由电子可在外部电路中产生流动。

10.1.5　水伏发电

不同于太阳能电池的光伏发电，水伏发电技术主要依赖于双电层中离子的动态分布[9]。如图 10.6 所示，在一个离子液滴沿固体表面的运动过程中，固液界面间的离子会重新分布，并在固液界面的前端和后端形成赝电容（分别表示为 C_F 和 C_R），从而实现充电/放电的过程。其中，对应于移动边界 L（dL/dt）模型的充电/放电过程所产生的电势，也被称为拉动势。以速度 v 在导电固体表面移动的离子液滴的等效电流可以表示为[10]：

$$I = \frac{dq_e}{dt} = \varphi \frac{dC_F}{dt} = -\varphi \frac{dC_R}{dt} = \varphi W C_0 \frac{dL}{dt}$$

式中，φ 是相对于斯特恩双电层（Stern 层）的表面电势；C_0 是每单位面积的赝电容；W 是液滴下的导电固体表面的宽度；L 是液滴行进的距离。

而对应于液滴撞击驻极体薄膜来发电的机制则源于电阻 - 电容器（RC）电路的液滴动力学，该种情况下的动态双电层电容以函数的形式随扩散液滴的导数面积（dS/dt）而变化，相应的感应电流可以被量化为[11,12]：

$$I = \frac{dq_e}{dt} = \varphi \frac{dS}{dt}$$

式中，φ 是固体的表面电荷密度；S 是液滴的面积。

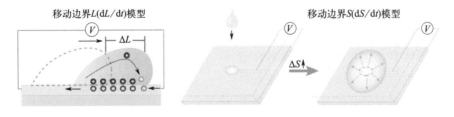

图 10.6　水伏发电的工作示意图[9]

10.1.6 磁电发电

磁电纳米发电机（magnetoelectric nanogenerator）的发电机理主要是基于电磁感应现象。该现象自 1831 年被法拉第发现以来，已逐步成为人类社会的主要供电方式，来自化石燃料、水、风能以及核能的能量均可通过电磁感应转化为电能[13]。然而以往的电磁感应设备往往会采用刚性磁体，这与水凝胶柔性传感器所需的机械鲁棒性是相矛盾的[14]。为此，研究人员将磁性粉末分散在聚合物弹性体中，利用嵌入在弹性体中的导电螺旋来实现各向异性的机电转换，使水凝胶传感器具有自供电功能（图 10.7）[13]。不同于传统的压电概念，施加在柔性传感器上的机械力可通过改变磁粉和导电螺旋的相对位置来影响基于磁电效应的电输出，输出电压可以表示为[13]：

$$E(V) = -N\frac{\Delta\Phi}{\Delta t} = -\sum_{i=1}^{N}\frac{\Delta\Phi_i}{\Delta t} = -\sum_{i=1}^{N}\frac{\Phi_i(\text{after}) - \Phi_i(\text{before})}{\Delta t}$$

式中，$E(V)$ 是输出电压；N 是螺旋层数；$\Delta\Phi$ 是总的磁通量变化，$\Delta\Phi_i$ 是每个等效环的磁通量变化；Δt 是复合弹性体的压缩时间。

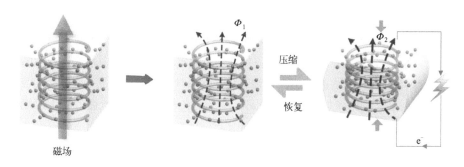

图 10.7 磁电发电的工作示意图[13]

10.1.7 混合发电

如前所述，基于单种纳米发电效应开发出的自供电水凝胶传感器往往存在输出电流低、输出电压低以及能量收集场景要求高的限制[3,15]。对于依靠太阳能结构发电用作传感系统电源的水凝胶传感器，它们的能源供应仅限于太阳光或环境光，一旦处于黑暗条件就会失去相应的功能。而基于机械能转化的 TENG/PENG 同样面临相应的困境，若该传感器附着的部位不能出现大幅度的动作变化，往往会出现能量收集过少而不足以充当能量供应装置的情形。为了解决这一问题，研究人员开始在水凝胶传感器中构建混合纳米发电结构（如摩擦-压电混合纳米发电、摩擦-压电-热释电混合纳米发电）以扩大其能量的获取范围，同时提高能量收集效率[16,17]。

10.2 自供能水凝胶传感器的典型应用

基于10.1节中所述能量收集技术，大量具有不同自供能机制的水凝胶传感器被开发出来，并被应用于力学传感、健康保健和环境监测等领域。这不仅避免了由定期更换电池所产生的额外经济负担和环境问题，还回收了原本无法继续利用的能量，有效地缓解了目前面临的能源危机和环境污染问题。

10.2.1 物理传感

水凝胶是由聚合物网络和水分子构成的，前者使水凝胶成为弹性体，而后者赋予了水凝胶离子导体的特性使其可以远距离传输高频电信号。高分子网络的网孔尺寸（约10nm）远大于水分子的尺寸，这使得水凝胶中的水分子具有和液态水相似的物理和化学性质[18]。因此可以通过往水凝胶中添加可溶性盐（LiCl，NaCl，KCl，$MgCl_2$，K_2SO_4等）来调节它的电导率。作为一种机械响应的黏性自供电水凝胶传感器，它能够感知到各种外部刺激（拉伸、压缩）并以电信号的形式进行反馈[19]。当水凝胶传感器被拉伸/压缩时，相应的电阻/电容会随着应变/压力向上非线性增加［图10.8（a）、(b)］。此外，自供电水凝胶传感器的黏附特性使得其可直接黏附到人体表皮组织上，并随着人体关节一起运动［图10.8（c）］。因而，该类传感器被广泛应用于呼吸、关节弯曲运动、喉咙振动以及脉搏等物理参数监测等生物力学传感［图10.8（d）］[20-26]。

基于自供电水凝胶传感器的上述物理传感特性，Guo等[27]进一步开发了一种可用于全方位婴儿运动监测并提供实时预警的基于TENG自供能的琼脂水凝胶传感器［图10.9（a）、(b)］。通过将该传感器贴在婴儿身体的不同部位从而构成一个区域传感器网络，以用于捕捉和收集婴儿身体不同部位特定运动的电脉冲信号。由于这些信号具有彼此不同的特性，可以利用深度学习算法来快速准确地实时识别婴儿的运动模式［图10.9（c）］，学习精度、学习率和损失函数均反映出该算法的高效性［图10.9（d）］。经过深度学习算法识别后的运动信号可以被无线传输到监护人的手机上以实现远程监护，该智能系统为物联网时代的婴儿护理提供了一个范例。

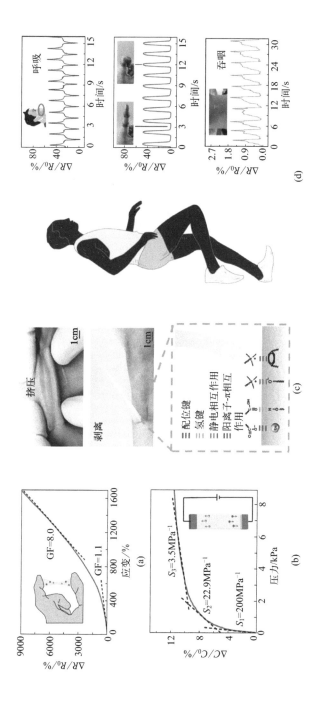

图10.8 （a）典型水凝胶传感器的应变依赖电阻变化；（b）典型水凝胶传感器的压力依赖性电容变化；（c）典型水凝胶传感器的机械顺应性和黏附于人体皮肤的特性；（d）用于生物力学传感的自供电水凝胶传感器[19]

第10章 仿生自供能水凝胶传感器 213

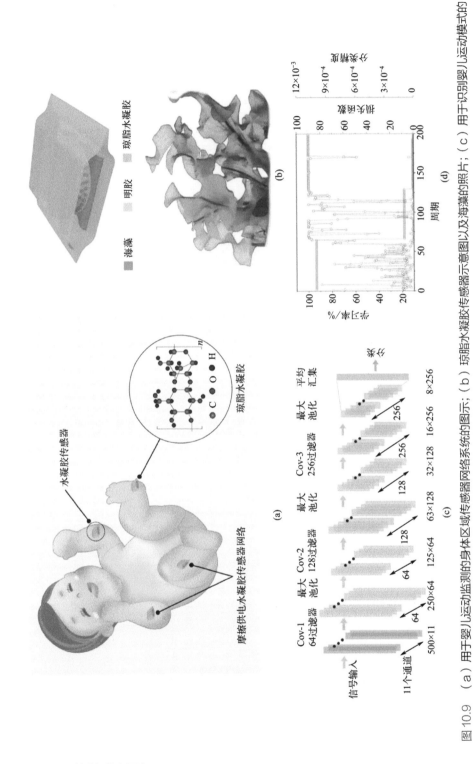

图 10.9 （a）用于婴儿运动监测的身体区域传感器网络系统的图示；（b）琼脂水凝胶传感器示意图以及海藻的照片；（c）用于识别婴儿运动模式的深度学习算法的架构；（d）深度学习算法在 200 个周期内的分类精度、学习率和损失函数[27]

214　仿生智能水凝胶

10.2.2 健康护理

水凝胶的离子电子学主要依赖于没有物质或电荷穿过界面的非法拉第过程[18]。而且，不同于液体水性电解质，水凝胶是固体，这使其可以直接连接到离子设备或活体组织上以传输相应的电信号[28]。因而，水凝胶传感器也一直被用于肌电图（EMG）、心电图（ECG）以及脑电图（EEG）等人体电生理信号的测量（图 10.10），这对于心脏、肌肉以及神经相关疾病的诊断和预测具有重要的意义[29-31]。

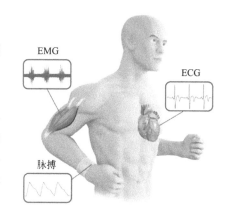

图 10.10 用于人体电生理信号监测的自供电水凝胶传感器[31]

除了用于上述心脏、肌肉或神经相关疾病的电生理信号的检测外，自供电水凝胶传感器还可通过对汗液中电解质或代谢物等可溶性生物标志物的检测，发现和预防囊性纤维化等疾病。Qin 等[32] 通过组装由聚二甲基硅氧烷（PDMS）封装的 TOCNF/PANI-PVAB 水凝胶和离子选择性膜（ISM），构建出了一种兼具优异拉伸性和自修复性的自供电水凝胶汗液传感器（图 10.11）。该水凝胶传感器能够以高选择性和灵敏度实时检测人体汗液中的 Na^+、K^+ 以及 Ca^{2+} 浓度以实现健康监测，这拓宽了自供电水凝胶传感器在健康监测领域的应用。

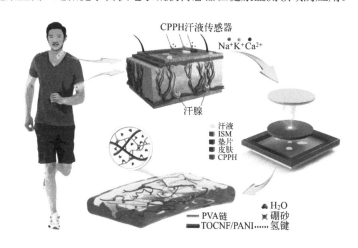

图 10.11 用于自供电汗液感应的纤维素基导电水凝胶及其工作示意图[32]

此外，患者、医护人员和医疗设备之间的人机交互（HMI）在智能医疗系统中起到了至关重要的作用。智能医疗监护系统不仅有利于满足患者的医疗需求，还能为医护人员提供相应的帮助［图 10.12（a）］。Yang 等[33] 开发了一种基于聚乙

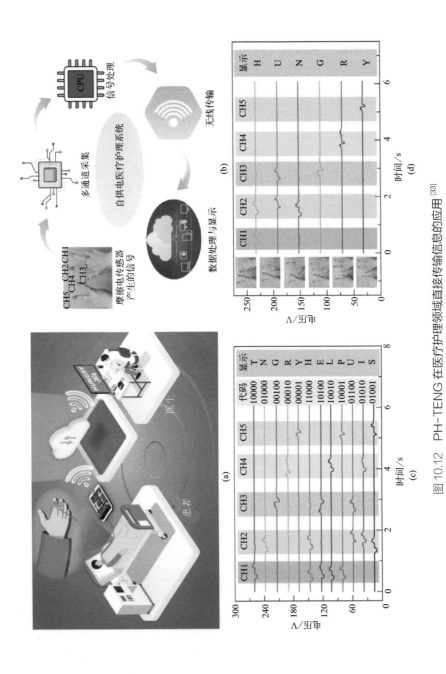

图 10.12 PH-TENG 在医疗护理领域直接传输信息的应用[33]

(a) 用于远程监控和传递患者帮助信息的自供电医疗护理系统示意图；(b) 自供电医疗护理系统中涉及的摩擦电传感器方案图；(c) 多个通道采集到的信号经过格雷码编码后定义为相应的字母；(d) 不同手势的组合表达"饥饿"(HUNGRY)

烯醇/植酸（PVA/PA）水凝胶的摩擦纳米发电传感器，该器件具有优异的力学和电学性能。通过将该传感器与患者的手指相结合［贴在患者手指上，图10.12（b）］即可构成一个微型传感器，当患者需要帮助时其只需弯曲自己的手指，相应的弯曲信号就会被基于LabVIEW的软件智能医疗监护系统所捕捉并识别，进而生成相应的求助信号传送给监护人［图10.12（c）、（d）］。该工作不仅展示了基于TENG自供电水凝胶传感器在医疗HMI领域的应用价值，还为自供电水凝胶传感器在脑机接口领域明确了发展方向。

10.2.3　环境监测

近年来，为了营造更好的生存环境，人类越来越注重与自然和谐共生的绿色发展。其中，温度、湿度和污染物水平是影响人类生存环境的关键因素，因此可通过实时监测这些参数来了解它们的变化规律，从而预防可能出现的环境污染、气候变化以及粮食安全问题。而自供能水凝胶传感器的出现，为我们以低碳绿色的手段大规模构建相关监测系统铺平了道路。

温度和湿度可以直接影响到我们对环境舒适度的感知，因此及时掌握相应参数的变化规律对预防极端气候的出现有着重要作用。然而当前市场上主流的温湿度传感器往往需要外部供电，这不利于低碳节能社会的构建。为了降低相应能源的消耗，Xia等[34]构建了一种基于水凝胶的梯度聚电解质膜（GPM）自供电水凝胶传感器，其供能机制主要依赖于GPM膜对离子选择性单向渗透过程中的化学能的收集。该传感器在水合情况下可以展现出基于热响应的自感应电位，而在脱水/干燥情况下则表现出对湿度敏感的跨膜电位，因而能可靠、精确地反馈出其所处环境的温度和湿度变化情况（图10.13），为未来开发具有多重传感功能的自供电离子感知系统提供了新思路。

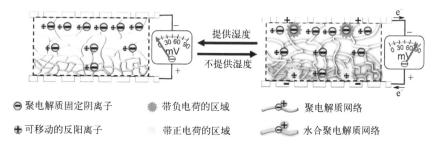

图10.13　用于环境温度和湿度传感的自供电水凝胶传感器[34]

农业生产过程中往往需要消耗大量的化学合成肥料，这不仅降低了相应的经济效益还增加了温室气体的排放量[35]。Hsu等[36]利用电感耦合等离子体蚀刻法构建的一种基于 PAA-RGO-PANI 双网络自供电水凝胶传感器的自供电的轻型施肥系统，通过水凝胶传感器中的摩擦纳米发电结构将环境中的机械能（气流、雨水、声音）转化为施肥系统可利用的电能以实时监测植物的生长情况和铵含量，并适时提供 LED 光照（"光肥料"）以促进植物的生长（图 10.14）。自供电水凝胶传感器在高效率和可持续发展的绿色智能农业系统的构建中展现出巨大的潜力。

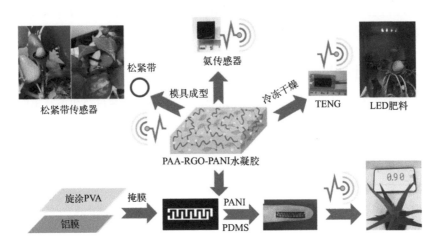

图 10.14　用于促进和监测智能农业中植物生长的自供电水凝胶传感器[36]

二氧化硫（SO_2）是一种对人体健康和生态环境均有着极大危害的气态污染物[37,38]，因此快速准确地检测出空气中 SO_2 的含量对于预防酸雨等二次污染具有重要的意义。然而，目前大多数 SO_2 传感器都需要外部供能，部分传感器甚至需要在高温条件下才能进行有效检测，导致相应的检测能耗非常高。对此，Wang 等[39] 开发了一种由海洋波浪能驱动的基于液-固 TENG 能量转化效应的自供能 MXene/TiO_2/SnSe 水凝胶传感器用于空气中 SO_2 气体含量的检测。如图 10.15

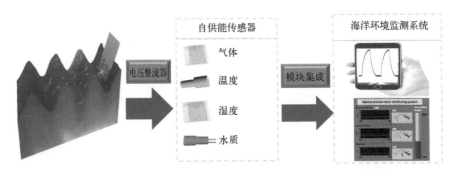

图 10.15　用于海洋环境监测的自供电水凝胶传感器[39]

所示，该传感器不仅能灵敏检测出海洋上空空气中 SO_2 的含量，还能同时检测空气的温度、湿度以及海水的水面高度；未来可进一步开发出用于检测海水水质（如重金属离子含量、海水 pH、N/P 元素的含量等）的自供能水凝胶传感器。

本章系统地介绍了纳米发电效应在自供电水凝胶传感器中的理论、最新发展和实际应用。纳米发电技术不仅免去了装置对外部电源的需求，同时还有效回收了原本无法继续利用的能量。所介绍的代表性例子——利用相应的纳米发电机将人体运动时产生的机械能／热能，汗水中的化学能，环境中的太阳能／磁能，风能、波浪能转换成相应的电能为水凝胶传感器供电，对本领域的未来发展具有很高的启发性。

参考文献

[1] Zhao L, Li H, Meng J, et al. The recent advances in self-powered medical information sensors. InfoMat, 2020, 2（1）: 212-234.

[2] Dong K, Peng X, Wang Z L. Fiber/fabric-based piezoelectric and triboelectric nanogenerators for flexible/stretchable and wearable electronics and artificial intelligence. Advanced Materials, 2020, 32（5）: 1902549.

[3] Wu Z, Cheng T, Wang Z L. Self-powered sensors and systems based on nanogenerators. Sensors, 2020, 20（10）: 2925.

[4] Xu F, Li X, Shi Y, et al. Recent developments for flexible pressure sensors: A review. Micromachines, 2018, 9（11）: 580.

[5] Wang Z L, Yang R, Zhou J, et al. Lateral nanowire/nanobelt based nanogenerators, piezotronics and piezo-phototronics. Materials Science and Engineering: R: Reports, 2010, 70（3-6）: 320-329.

[6] Yang R, Qin Y, Dai L, et al. Power generation with laterally packaged piezoelectric fine wires. Nature nanotechnology, 2009, 4（1）: 34-39.

[7] Crispin D X. Retracted article: towards polymer-based organic thermoelectric generators. Energy & Environmental Science, 2012.

[8] Tian B, Zheng X, Kempa T J, et al. Coaxial silicon nanowires as solar cells and nanoelectronic power sources. Nature, 2007, 449（7164）: 885-889.

[9] Wang X, Lin F, Wang X, et al. Hydrovoltaic technology: from mechanism to applications. Chemical Society Reviews, 2022.

[10] Yin J, Li X, Yu J, et al. Generating electricity by moving a droplet of ionic liquid along graphene. Nature nanotechnology, 2014, 9（5）: 378-383.

[11] Wang X, Fang S, Tan J, et al. Dynamics for droplet-based electricity generators. Nano Energy, 2021, 80: 105558.

[12] Wu H, Mendel N, van den Ende D, et al. Energy harvesting from drops impacting onto charged surfaces. Physical review letters, 2020, 125（7）: 078301.

[13] Zhang X, Ai J, Ma Z, et al. Magnetoelectric soft composites with a self-powered tactile sensing capacity. Nano Energy, 2020, 69: 104391.

[14] Wang C, Wang C, Huang Z, et al. Materials and structures toward soft electronics. Advanced Materials, 2018, 30 (50): 1801368.

[15] Hu H, Zhang F. Rational design of self-powered sensors with polymer nanocomposites for human-machine interaction. Chinese Journal of Aeronautics, 2022.

[16] Chen X, Song Y, Su Z, et al. Flexible fiber-based hybrid nanogenerator for biomechanical energy harvesting and physiological monitoring. Nano Energy, 2017, 38: 43-50.

[17] Sun J G, Yang T N, Wang C Y, et al. A flexible transparent one-structure tribo-piezo-pyroelectric hybrid energy generator based on bio-inspired silver nanowires network for biomechanical energy harvesting and physiological monitoring. Nano Energy, 2018, 48: 383-390.

[18] Yang C, Suo Z. Hydrogel ionotronics. Nature Reviews Materials, 2018, 3 (6): 125-142.

[19] Zhang W, Wu B, Sun S, et al. Skin-like mechanoresponsive self-healing ionic elastomer from supramolecular zwitterionic network. Nature Communications, 2021, 12 (1): 1-12.

[20] Chen J, Zhang L, Tu Y, et al. Wearable self-powered human motion sensors based on highly stretchable quasi-solid state hydrogel. Nano Energy, 2021, 88: 106272.

[21] Hu K, He P, Zhao Z, et al. Nature-inspired self-powered cellulose nanofibrils hydrogels with high sensitivity and mechanical adaptability. Carbohydrate Polymers, 2021, 264: 117995.

[22] Ma C, Xie F, Wei L, et al. All-Starch-Based Hydrogel for Flexible Electronics: Strain-Sensitive Batteries and Self-Powered Sensors. ACS Sustainable Chemistry & Engineering, 2022.

[23] Luo X, Zhu L, Wang Y C, et al. A flexible multifunctional triboelectric nanogenerator based on MXene/PVA hydrogel. Advanced Functional Materials, 2021, 31 (38): 2104928.

[24] Liu Y, Wong T H, Huang X, et al. Skin-integrated, Stretchable, Transparent Triboelectric Nanogenerators Based on Ion-conducting Hydrogel for Energy Harvesting and Tactile Sensing. Nano Energy, 2022: 107442.

[25] Feng Y, Yu J, Sun D, et al. Extreme environment-adaptable and fast self-healable eutectogel triboelectric nanogenerator for energy harvesting and self-powered sensing. Nano Energy, 2022, 98: 107284.

[26] Chen C, Li Y, Qian C, et al. Carboxymethyl cellulose assisted PEDOT in polyacrylamide hydrogel for high performance supercapacitors and self-powered sensing system. European Polymer Journal, 2022, 179: 111563.

[27] Guo R, Fang Y, Wang Z, et al. Deep Learning Assisted Body Area Triboelectric Hydrogel Sensor Network for Infant Care. Advanced Functional Materials, 2022, 32 (35): 2204803.

[28] Bard A J, Inzelt G, Scholz F. Electrochemical dictionary. Springer, 2012.

[29] Wang Q, Pan X, Guo J, et al. Lignin and cellulose derivatives-induced hydrogel with asymmetrical adhesion, strength, and electriferous properties for wearable bioelectrodes and self-powered sensors. Chemical Engineering Journal, 2021, 414: 128903.

[30] Kim J N, Lee J, Lee H, et al. Stretchable and self-healable catechol-chitosan-diatom hydrogel for triboelectric generator and self-powered tremor sensor targeting at Parkinson disease. Nano Energy, 2021, 82: 105705.

[31] Li M, Zhang Y, Lian L, et al. Flexible Accelerated-Wound-Healing Antibacterial MXene-Based Epidermic Sensor for Intelligent Wearable Human-Machine Interaction. Advanced Functional Materials,

2022：2208141.

[32] Qin Y., Mo J, Liu Y, et al. Stretchable Triboelectric Self-Powered Sweat Sensor Fabricated from Self-Healing Nanocellulose Hydrogels. Advanced Functional Materials，2022：2201846.

[33] Yang J, An J, Sun Y, et al. Transparent self-powered triboelectric sensor based on PVA/PA hydrogel for promoting human-machine interaction in nursing and patient safety. Nano Energy, 2022, 97: 107199.

[34] Xia M，Pan N，Zhang C, et al. Self-Powered Multifunction Ionic Skins Based on Gradient Polyelectrolyte Hydrogels. ACS nano，2022，16（3）：4714-4725.

[35] Elbana T，Gaber H M，Kishk F M. Soil chemical pollution and sustainable agriculture. In The soils of Egypt，Springer，2019：187-200.

[36] Hsu H H, Zhang X, Xu K, et al. Self-powered and plant-wearable hydrogel as LED power supply and sensor for promoting and monitoring plant growth in smart farming. Chemical Engineering Journal, 2021，422：129499.

[37] Zhang L，Liu W，Hou K，et al. Air pollution-induced missed abortion risk for pregnancies. Nature sustainability，2019，2（11）：1011-1017.

[38] Tchalala M，Bhatt P，Chappanda K，et al. Fluorinated MOF platform for selective removal and sensing of SO_2 from flue gas and air. Nature communications，2019，10（1）：1-10.

[39] Wang D, Zhang D, Tang M, et al. Ethylene chlorotrifluoroethylene/hydrogel-based liquid-solid triboelectric nanogenerator driven self-powered MXene-based sensor system for marine environmental monitoring. Nano Energy，2022，100：107509.

第11章
总结与展望

11.1 仿生智能水凝胶软执行器
11.2 自供能水凝胶传感器

水凝胶代表一类重要的材料，具有水环境和广泛可调的物理化学性质。在过去十年中，致力于工程化具有增强性能的水凝胶的努力扩大了它们在众多应用中的机会，包括生物医学、软电子、传感器和执行器。水凝胶的力学设计使其变得坚硬，同时保持高含水量。自我修复机制和动态调节已被纳入水凝胶系统中，以随着时间的推移实现对其行为的控制。先进的生物制造技术进一步提高了我们构建复杂水凝胶结构的能力，这些结构具有跨多个长度尺度的分层组装结构。然而，仿生智能水凝胶领域的未来发展仍然面临挑战。

11.1 仿生智能水凝胶软执行器

在本书的第 9 章中，我们系统地介绍了基于仿生系统的水凝胶致动器的最新研究。所制备的水凝胶材料或器件虽然具有各自的独特优势，但仍有很大的改进空间。我们相信未来的研究将朝着赋予材料可编程可控响应性、提高材料耐久性以及功能进一步系统化的方向发展，如图 11.1 所示。

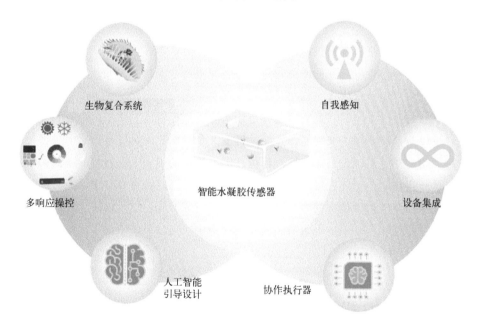

图 11.1 制备智能和自主水凝胶致动器的未来方向

（1）赋予水凝胶致动器可编程的响应能力

随着对多功能执行器的研究兴趣的增加，越来越多的研究将集中在响应性水凝胶上，因为它们可以更好地结合温度、光、电和磁等不同刺激-响应模式。然而，水凝胶致动器的开发和应用仍然受到各种刺激实施的限制。为保证刺激的可靠性和可控性，水凝胶致动器的致动和释放需要快速精确的实现。计算机编程的进步应该能够实现更精确的控制。目前我们对材料特性进行编程和控制外部刺激的能力的局限性限制了实际应用，未来将优化算法和人工智能（AI）应用于材料设计以及外部刺激的控制会是发展的重点。以水凝胶履带为例，为了达到对复杂运动的高精度控制，需要对材料的理化特性进行精心设计，并对外部条件进行精确操控。如果没有基于计算机控制和人工智能的新技术的支持，这可能很难实现。

(2) 提高水凝胶致动器的耐久性

目前的研究更多集中于提高水凝胶的力学性能，缺乏对其耐久性的研究，而耐久性往往是实际应用中面临的关键问题。由于内部水分的蒸发，水凝胶在开放系统中通常会脱水，导致材料的柔韧性、透明度、导电性、响应性下降。尽管有部分研究已经开始着手解决这些问题，但情况并不理想。另外，水凝胶在液体环境中的吸水膨胀问题多年来一直困扰着许多研究人员，相关的研究仍有改进的空间。最后，应对极端条件的能力是另一个重要问题，因为水凝胶致动器需要在可能具有各种极端温度/压力条件的实际场景中使用。因此，水凝胶致动器在这些条件下运行的能力将进一步拓宽其应用范围。

(3) 进一步系统化

从系统的角度来看，未来的趋势是让水凝胶执行器自主感知、响应和完成编程任务，从而使系统更加智能。然而，目前大多数水凝胶执行器仍处于开环控制阶段。机器学习可以帮助实现这一目标，因为从自动驾驶汽车到智能建筑的不同领域的自主控制，已经在机器学习的帮助下实现。因此，将具有学习能力的控制系统集成到智能水凝胶机器中是一个很重要的研究方向，它可以将一般的非线性传感信号映射到控制系统中水凝胶的驱动命令中。在该系统中，水凝胶传感器可以从闭环操作系统获得反馈并自主调整其行为。

11.2 自供能水凝胶传感器

在迈向高度集成和智能化的物联网发展趋势中，传感器作为信息收集转换的开端，普遍需要消耗大量能源。在绿色发展的趋势下，传感器朝着低功耗甚至自供能的方向发展，并在消费电子和医疗保健电子领域迅速兴起。但是，规模化的生产和应用仍然存在一些挑战和新技术的发展空间（图 11.2）。

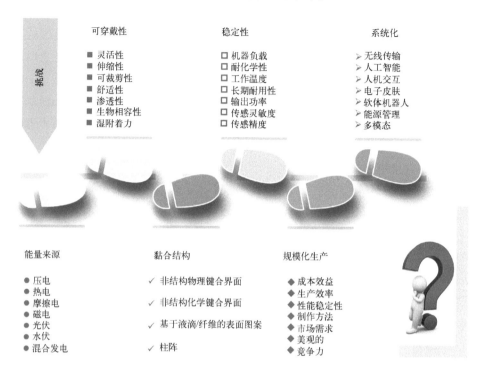

图11.2 基于纳米发电机的自供电水凝胶传感器的发展路线图

首先，大多数自供电水凝胶传感器面临的挑战是其能量获取环境/条件单一，使得其应用范围受到了严重的限制。虽然在水凝胶传感器中构建混合发电装置是目前的主流发展趋势，但同样存在其局限性。在现有的水凝胶传感器中，不同发电效应的组合往往是简单的堆叠，这虽然提高了整体的能量收集能力，但单个纳米发电机的能量收集效率却会降低。此外，多个纳米发电效应的叠加会使得所用材料更多，结构更为复杂，不利于整个自供电水凝胶传感器的微型化构建。因此，如何通过调整纳米发电结构的化学成分来降低发电装置所需的空间，以结合高能量收集效率和广阔的应用场景是研究人员迫切需要解决的问题。

其次，对应用于健康监测领域的自供电水凝胶传感器，要兼顾其附带属性的构建，如生物兼容性、可植入性、湿黏附性、拉伸性、灵活性、可降解性等。根据自供电水凝胶传感器在人体附着位置的不同，其应用场景可分为体内传感器和体外传感器。

前者通常用于采集体内各器官的生物信息，以用于监测用户的健康状况。例如，心内膜压和心律失常等医学信号可以有效为心室颤动、室性早搏等心血管疾病的发现提供及时的线索。一般而言，这一类自供电传感器是需要兼具可植入性和生物相容性等特点。但与体外环境相比，体内场景复杂且易受周围组织的影响，因此如何有效提升体内自供电水凝胶传感器的生物安全性、抗干扰能力以及传感器的微型化是目前面临的主要挑战。此外，体内自供电水凝胶传感器在生物医学或植入式的健康治疗商业化应用还受到其生物降解性的限制，因此如何在保留其相应功能的前提下用生物降解性材料来构建自供电水凝胶传感器也是未来研究中需要解决的问题。

体外自供电水凝胶传感器则主要用于收集汗水、脉搏、体温、呼吸、声音以及肢体运动等信号，该类自供电传感器通常需要兼具可拉伸性和灵活性。由于长时间暴露在空气中以及在监测肢体运动时会出现重复性的机械运动，因此水凝胶传感器的保湿性、结构稳定性以及电信号稳定性是需要考虑的重要因素。水凝胶在潮湿表面上的弱且不稳定的黏附会导致界面失效和随后的功能丧失，这严重阻碍了它们的功能和应用。因此，不论是体内传感器还是体外传感器，传感器的强湿界面黏附特性都是研究人员在设计时需要考虑的一个重点。

最后，应进一步研究和提高自供电水凝胶传感器的稳定性或长期耐用性。在实际应用中，水凝胶传感器会受到器件结构、组件材料的机械强度、摩擦层材料的磨损、电极之间的弱连接以及环境中的水分和温度等因素的限制。这需要研究人员通过传感器结构设计/部件的优化，新型摩擦层高分子材料的使用，强界面黏合结构的设计，以及多预设环境下的测试来进一步改进并提升相应传感器的使用稳定性和寿命。例如，对于植入体内的自供电水凝胶传感器，我们可以采用防水聚合物薄膜对整个传感器进行外部封装或在最外层镀上一层防水高分子涂层以减轻周围组织长期包裹带来的腐蚀和溶解等影响，从而提升其使用寿命。而对应用于运动监测的自供电水凝胶传感器，则应侧重于分子链之间的化学键的构建以提升水凝胶传感器的机械强度，从而使其可长时间稳定应用于往复性机械运动的检测。若该水凝胶传感器是长期使用于空气环境或低温环境中的，则需着眼于自供电水凝胶传感器内部的保湿结构和抗冻结构的设计，以提升其性能稳定性和使用寿命。

随着越来越多地采用跨学科的方法，可以合理地结合多种策略，从而创造出具有增强性能的水凝胶。在多尺度上组装多成分的材料设计可以利用协同效应并促进新功能。由于多尺度成分之间的特定排列或相互作用，可能会出现属性组合，从而为水凝胶的进一步创新提供机会。此外，水凝胶功能器件的整体性能取决于不同组件之间的协同作用。因此，有必要开发一种综合方法，结合新型智能水凝胶致动器设计中涉及的所有方面。可以预见，将技术应用与化学设计、人工智能和机器学习相结合具有吸引力。总而言之，智能水凝胶致动器的开发是一项长期的跨学科研究工作，需要来自化学、材料、电气工程或计算机科学等不同领域的研究人员通力合作。